Gold Gamble

by

Roberta Martin Starry

Roberta Martin Starry

Book design by William Erwin
Cover illustration by Danny Jones
Production by BookMaster's Inc.

Originally published in 1974 by the Maturango Museum as MATURANGO MUSEUM PUBLICATION 7

Engler Publishing is an imprint of George N. Engler & Associates

Printed in the United States of America

Engler Publishing
George N. Engler & Associates
P.O. Box 6311
Woodland Hills, CA 91365-6311

Library of Congress Catalog Card Number: 00-110158
International Standard Book Number: 0967463718

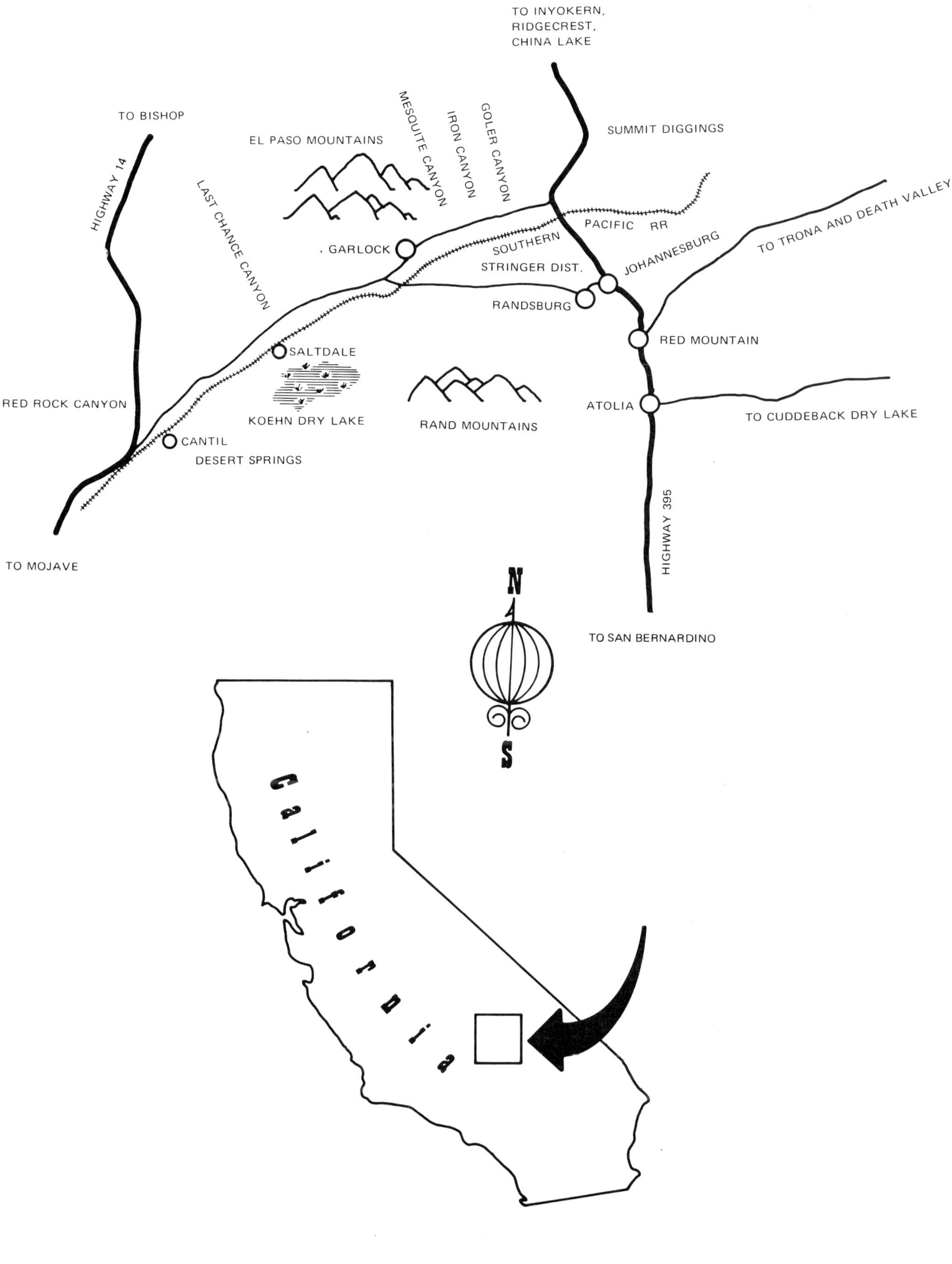
TO INYOKERN, RIDGECREST, CHINA LAKE
TO BISHOP
SUMMIT DIGGINGS
EL PASO MOUNTAINS
MESQUITE CANYON
IRON CANYON
GOLER CANYON
HIGHWAY 14
LAST CHANCE CANYON
PACIFIC RR
SOUTHERN
TO TRONA AND DEATH VALLEY
GARLOCK
JOHANNESBURG
STRINGER DIST.
RANDSBURG
RED MOUNTAIN
SALTDALE
RED ROCK CANYON
KOEHN DRY LAKE
RAND MOUNTAINS
ATOLIA
TO CUDDEBACK DRY LAKE
CANTIL
DESERT SPRINGS
HIGHWAY 395
TO MOJAVE
N
S
TO SAN BERNARDINO
California

CONTENTS

Old freight route from Johannesburg to Ballarat and the Panamint Mountains, 1896. Pilot Knob on left was landmark seen for miles and mentioned by early explorers. *(Author's collection)*

FOREWORD

Two great waves of interest have swept over us in the past 20 years. One is an intense fascination with deserts, including the great Mojave Desert of California. The other has to do with that part of the past that is almost reachable to us. It is a nostalgia for days grandmother and grandfather knew—those days when the West was still pretty wild, all skies were very blue, and gold was plentiful in some remote areas.

Gold Gamble satiates both these urges. The ghost town of today, as starkly desert as ever, comes alive with the fascinating real people who prospected, who struck it rich, who supplied the area, and filled its busy streets. The photographs and the anecdotes show what it looked like and how the townsfolk reacted. And the fact that Randsburg, Johannesburg, Red Mountain, and Garlock can be visited today, with many a head frame, slag pile, miner's shack, and a couple of long bars still to be seen, adds zest for those who have been there or for those who would like to go. Garlock is the site with fewest visible reminders; it had the least permanent structures and is fast disappearing.

The desert is harsher on its scars—already placered parts of the Mother Lode country have been softened by water action and vegetation. A prospector's glory hole in the Rand District may look as if the miner had just put down his pick or his single jack. The raw barren hills seem too massive and unsculptured to be hiding precious metals.

Yet, here it was, gold and silver for the finding, fortunes to be made and lost, hardy men and women, towns to be built on barren hillsides. Always—there must be water, as true today as when the towns were built.

Roberta Starry has written with insight into the lively days of the desert bonanza. Read along and feed your nostalgia with vivid simple tales and vivid memories of the tough men and women who conquered, for a time, that part of the desert known as The Rand.

The Maturango Museum is both pleased and proud to bring you this lively record of a little known part of the mining history of California.

K. H. Robinson
Director, Maturango Museum

ACKNOWLEDGMENT

Special thanks to the countless men and women, most of whom are now gone, who paused to share their memories with us. Like the making of a tapestry, it was their reliving of a place and time that gave color to otherwise dull facts. Each life, a thread running through a pattern of hardships, good times, sadness and laughter, gives a picture of the gold epic as played out in the Rand area.

To those who devoted many hours to the preparation of this book we express deep appreciation: William Erwin, book design; Danny Jones, cover illustration; Thomas Boyd, production; Maxine Malone, Maureen Donohue, Helen Deis, Barbara Manley, and Elizabeth Taylor, composition and proofreading; and Florence Dinsmore, editor.

Tracks in the Desert

As the 1840s came to a close, gold in California drew pioneers and immigrants like a great magnet. Yet comparatively few moved along the inland slope of the Sierra, and even a smaller number ventured into the arid expanse of the Mojave Desert.

But here, in the Fremont and Golden Valleys and in the canyons and washes of the El Paso and Rand Mountains, seemingly a world apart from the coast and the gold fields of the north, a remarkable treasure was waiting to be discovered. In this eastern corner of Kern County and a portion of San Bernardino County, the area to become known as the Rand Mining District would be the site of a tremendous gold mining boom by the late 1890s. But until then, the isolation of the desert, the searing heat and driving sandstorms, and the rugged mountain slopes rebuffed all but the sturdiest, the roughest and the toughest who dared to venture into the area.

News traveled slowly. Mexican riders carrying letters and messages in leather pouches slung across their saddles were developing a courier system, but few reached the less settled regions. Pack trains of burros and mules carried the burden of merchandise and mail, each animal carefully working his way past projecting boulders, along narrow ledges, and through the hot, dry sand with his 200- to 400-pound load. Delivery rates varied, and a schedule was virtually unheard of, but for from forty to sixty cents, a letter would eventually reach the desert areas east of the Sierra.

Those enterprising persons who tried to develop a more systematic communication and delivery service invariably faced defeat at the hands of the lawless. Civil government and laws were almost nonexistent in California, as were the means of enforcement. Crime was not deterred by punishment, and murder carried a lighter sentence than theft. In 1850 it was estimated that approximately 1,000 men throughout California were making a lucrative living as thieves and murderers. Bandits, army deserters, highwaymen, and criminals of all description endangered the life of the average citizen. Travelers were lassoed from their horses and left to die or struggle on foot to reach a settlement before strength and water gave out. The theft of cattle and horses became so prevalent that their value diminished and ownership became a drawback rather than an advantage.

Life was heavy with struggle and uncertainty; respected people became infected with the atmosphere of crime, taking what they could get with little regard for others. Fremont and Golden Valleys became infected, too, for the high grass was good grazing for animals, wild game fed the men, abundant springs provided water, the valley floor was easy to travel, and the mountains provided deep canyons that were ideal for hideouts. Here was an ever-present threat to the unwary traveler or potential settler.

A gold claim notice in the El Paso Mountains, dated 1853 and filed in the name of Herman Johnson, is the first known document to verify the presence of prospectors in the area that was destined to become the wealth-producing mining district of central California. By 1894 when W. J. Langdon discovered that early claim notice, there were small settlements and numerous mining projects throughout the mountains and valley. But Langdon's find did not result in locating a lost mine or a producing mine; he had only a rusty gold scale and a whiskey bottle containing six dollars' worth of gold dust, plus the mine claim papers.

A survey of the area, made in 1854, contained no mention of mines, but the omission is understandable since the surveyor claimed the mountains were too steep and impassable. Field notes in longhand, presently preserved in the Kern County Surveyor's office, give a comprehensive picture of the surveyor's impressions.

The document was made by W. S. Washburn, Deputy Surveyor under special instructions, dated August 25, 1854: by I. G. McDonald, Deputy Surveyor, dated March 17, 1855; and J. A. Tiny, Surveyor, dated March 19, 1855. The notes not only contain footage measurements, but indicate Indian trails, where dry creeks were in evidence, and the location of active springs. The surveyors camped at Mesquite Springs, and prepared charred mesquite wood for marker posts. The "charred stick was placed in a mound of earth with a trench and pits for corner sections." This method of marking lasted well in the East and Midwest, but was a complete failure on the desert where winds moved the mound of earth, toppled the stick, and then blew it away.

Washburn's notes deal with the area that in time became Garlock and nearby homestead lands:

> The quality of the land is considerably below the common average. It is generally very sandy, covered with greasewood brush, although it has an abundance of fine bunch grass which I think superior to any grass that I have ever seen to fatten an animal quick. This township has one spring affording a considerable quantity of water though it is slightly salty.
>
> *August 25, 1854:* Had to stop because of heat and had to carry water.

The honesty of a survey made by W. W. Carlton, December 14 through 19, 1876, is questioned by present surveyors, who point out that even with many teams working, they could not have covered the territory that was to have been surveyed for homesteaders. For two dollars per mile, they were required to dig trenches and char the posts used as markers. No indication of trench or post has been found, but lengthy reports were filed.

The field notes are interesting: "Set a greasewood post 3 inches in diameter." Three inches would be immense for greasewood, which is noted for its slender, flexible branches that start at ground level and form a bush without a large central trunk. The notes continued: "Dug a trench 5 foot square and 1 foot deep and 1 foot square, 18 inches from the outside of the sides of the trenches." Such a trench would be slow digging in the adobe of the foothills.

In a summary, "General Description," the survey reads:

> The southern portion has been surveyed many years and being a portion of the Mojave plains, is somewhat sandy and gravelly with considerable sage. There is however a good growth of grass which gives an abundance of feed for stock. The northern portion is somewhat broken and hilly. But the soil is generally of better quality and more luxurious grass than on the plains. Sufficient water is found for purposes of irrigation by water from the mountains; the land can be made productive.

Carlton's description was the basis of many a homesteader's hopes through the coming years; apparently few noted that it had been written in the winter when rains and cool weather changed the face of the desert.

California's growth between 1860 and 1880 was slow; the desert was still a remote land, lacking good connections with the rest of the state and nation. Travel remained slow, expensive, and dangerous. In 1862 the greatest flood known to that date spread from Shasta to the Mexican border and was followed by nine years of severe drought. Times were hard, and those who did migrate into the more arid regions had little or no money.

Nevertheless the inland country was moving ahead. Cattlemen and prospectors were becoming more familiar with the Fremont and Golden Valleys, the landmarks like Pilot Knob, the springs, and the canyons. Land use was assuming importance, although the cattlemen usually kept to the valley where feed was abundant, while the prospectors in increasing numbers worked the rugged mountains and canyons.

By this time, mining interests in northern California were in a slump. Scattered discoveries in southern counties revived hope that there would be another gold rush. W. H. "Hamp" Williams, whose son by the same name later played an important role in the Rand area, discovered a good deposit six miles west of the South Fork of the Kern River. The find ran $200 per ton, milled by the primitive Mexican rock drag, arrastra process. News of the discovery traveled unbelievably fast, and within the month over a hundred claims had been filed. Farther to the south on Lytle Creek in San Bernardino County reports of nuggets the size of hen's eggs created a small gold rush there.

In 1861 the Milligan Mine, in the El Paso Mountains of eastern Kern County, had a shaft to a depth of 175 feet in a ledge that produced gold worth from $50 to $150 a ton of ore at a time when gold was bringing anything the miner could get, up to $22 an ounce. Fact and rumor of gold finds, no matter where they happened, set in motion an influx of miners hoping for a new chance at wealth. Young men, who felt cheated at having been too young to get in on other strikes, grasped at the opportunity a fresh discovery offered. Danger of outlaw, Indian, or summer heat was of little concern when gold was in the balance.

The freighting of supplies to the growing mining population was becoming big business in spite of the expense and the time involved. The great feat of transporting thousands of tons of supplies was accomplished mainly by means of huge wagons. H. H. Bancroft described them as averaging 6 feet in depth of hold and 17 feet in length. Cost ranged from $1,500 to $8,000. With a harness costing from $300 to $600 and mules $500 to $1,000 a pair, the price of an outfit often exceeded $5,000.

One of the established routes into the mines being developed in the Panamint Mountains and along the Slate Range skirted the foothills of the El Paso Range, with El Paso City and El Paso Well serving as supply points and watering spots. Ore was freighted out; food, feed, liquor, and mining equipment were brought in on the return trip. As demand for supplies increased, new way stations developed and facilities at existing stops were improved, but it was still a long way between points. There were so few settlers that murder and banditry could continue to flourish with little fear of detection.

Few incidents received publicity beyond the local area. After a story had been repeated a number of times, passed on from one freighter to another, it often had little similarity to the real event by the time it reached the Los Angeles or San Francisco papers. The *Los Angeles Tri-Weekly* did note a crime committed in Fremont Valley in 1864. The owner of the Yarbrough Gold and Silver Company in the El Paso Mountains was murdered while waiting at Mesquite Springs with a shipment of ore or gold concentrate to be picked up by a freight wagon en route from the Slate Range to Los Angeles. Whenever such an incident occurred, the immediate cry was "Indians," without regard to evidence. The *Tri-Weekly* reported it as "assumed to have been Indians." Too often renegades, white and Mexican, got away with the loot and let the blame fall on the innocent.

Shortly after the Yarbrough murder, the miners at the Manzanillo and Ophir Mines, also in the El Paso Range, were allegedly driven off by Indians, leaving 40 tons of gold ore on the dump. "Unknown" parties carried off the ore and realized $400 a ton. No one knows how the rumor started that anyone, Indian or otherwise, got $400 a ton.

Even though way stations were only 10 to 15 miles apart, depending on the difficulty of travel, prospectors back in the canyons passed days without sight of another human being. There was no one to talk to except the pack burro, the lizards scooting among the rocks, or the great open spaces that could not answer back. The lonely spot had its own sounds, however: desert winds whipping up a draw or playing in and out of the scrub brush were far from silent. Overhead was the whir of wings as the golden eagle, raven, or hawk circled or rode the air waves. At night the desert fox sent his light bark out over the land, joined by the coyote's quavering song. When solitude grew heavy or supplies ran out, the miner and prospector could find companionship at El Paso City, Mesquite Springs, Harry Ball's Station at Desert Springs, or the Red Rock Station. The population might be very small at any of these stations, but the sound of another voice, limited supplies, occasional mail, a possible poker game, and liquid refreshment could be found there.

Teamsters who made the long hauls spent all their daylight hours jolting along on the heavy wagons, beat by cold winter winds or baked in the sun of summer. The beautiful, calm, comfortable days of spring and fall helped offset the uncomfortable ones, but did nothing for the aloneness. There were only the mules to yell at or talk to. An occasional lizard, rabbit, or snake crossing the dust trail broke the hours of monotony by offering a moving target for the long whip. Guns and powder were used for necessity only; they were not to be wasted.

The creak and groan of the wagons, the squeak of leather, the sound of plodding feet, and the rhythm of the lead team's bells were hypnotic. Drivers told of sleeping for miles, lulled by the repetitive sounds and the sheer monotony of level valleys. But rough areas, sand, and steep grades found them awake and capable of meeting any problem. They were held in high esteem by other men for their ability to handle any number of teams and their seeming lack of fear on the dangerous stretches.

Picturesque Red Rock Canyon became a familiar name to all freighters in and out of southern California as drivers exchanged stories and told of the problem areas on their routes. Remarks about the beauty and unusual rock formations were overshadowed by descriptions of the deep sand, hidden boulders, and the ever-present danger of flash floods.

Remi Nadeau in *City Builders* tells of his great-great-grandfather freighting from Cerro Gordo through Red Rock Canyon. Nadeau had agreed to haul ore to Los Angeles only on the condition that funds were available to set up adequate stations along the way; Red Rock was to be one of the improved outposts. His freighting was unique because he had 80 teams on the road between Los Angeles and Cerro Gordo, operating on an hourly schedule like a stage line. Fresh teams took a load of bullion to the next station, then returned to their home base with a wagon loaded with supplies of all kinds, with most of the space given to food, hay, and grain.

The deep sand of Red Rock Canyon pulled the massive freight wagon wheels down until the wagon box dragged on the surface, and Nadeau found it advisable to double his teams to 20 mules in order to move the heavy loads of ore or silver ingots to the Los Angeles market. Nadeau and other freighting companies also found it wise to have two wagons traveling through the canyon at the same time so that one could help the other out of sand or mud or over steep grades. Most to be dreaded in the canyon were wind and rain. Wind gusts were so vicious at times that wagons, loaded high, especially with hay, were blown over. Rain clouds were a warning to stay out of the canyon because sudden downpours created immediate flooding.

Present-day treasure hunters still search Red Rock Canyon for the silver bullion lost in a flood in the late 1870s. The shipment of silver bars, 2 feet long and 4 to 5 inches wide filled two freight wagons. The driver saw the storm clouds but believed that no rain was imminent; the 16-mule team, two wagons, and a water wagon were almost out of the canyon when he heard an unusual roaring sound. Looking around to see where the sound was coming from, he saw a 20-foot wall of water rolling down upon him. He jumped from the wagon, whipped the animals into a run, and then hurried to make his own escape up the side of a cliff.

Mules and wagons disappeared under the muddy, racing water, and though a search was made, no sign of wagons or silver was found in the flood's debris. In 1906 Rolo Miller, surveyor for the Los Angeles aqueduct, found two bars of silver in Red Rock that were believed to be part of that shipment.

A freighter known only as Big Jim, traveled the route along the eastern foothills of the El Paso Range in 1886. He chose to avoid the Red Rock route when hauling supplies to the Panamint mine camps: "I never went through that canyon that I didn't have trouble or have to help someone else out of a bad spot." Big Jim added that he looked forward to the overnight stop at El Paso City or Cow Wells, names he used interchangeably. That they were the same place is borne out by an 1880 township and census map that places El Paso City where Cow Wells and now Garlock appear on present maps. W. W. Elliott's history of San Bernardino County of 1882 also shows the eastern section of the adjoining Kern County, with El Paso City, Mesquite Canyon, and the dry lake along the freight route in their present positions.

Although Big Jim talked of El Paso City as having quite a few adults and children, the census of 1880 indicates a population of less than 100 persons. A one-room wooden store carried supplies for prospector, cowboy, and teamster. One woman in the settlement served meals in her one-room tent and lean-to shack; Big Jim liked her cooking and enjoyed talking with her prospector husband.

It was not unusual for the freighters to sleep and cook at their wagons. However, if they had been paid recently and were near good home cooking, they gladly passed up their own culinary skills. And considering the limited means of communication, it was remarkable how word got around of a gold strike or of a way station where a woman set a good table.

The name of Cow Wells developed probably because of the increased use of the area by cattlemen. Herders lived in tents or half-tents, with boards or rock completing the structure. Wells were dug by hand, and water was pulled up in buckets hand-over-hand by the cowboys.

Early in the 1890s cattlemen's names began to appear in records of the area. Stock from the Greenhorn and Tehachapi Mountains were brought into Fremont Valley when snow covered the higher country. William Landers, who ran great herds through the area, is credited with having his cowboys dig several wells in the vicinity of Cow Wells. His brand, an *L* far down on the leg, nearly to the hock, could be found from the Tejon Ranch on the south, through to the Fremont and Golden Valleys, east to the Mojave River, and north to the meadows of the Sierra.

William Cuddeback dug a well with a hand auger out on the dry lake that later bore his name and that still later became part of the Rand area. When water was reached just below the 40-foot level, a square casing of redwood was lowered, and a man known as Lovejoy built a water trough out of 2 by 12s. Sam Cuddeback, as a small boy accompanying his father on cattle drives, watched the well being dug. "The filaree and grasses grew high and abundant so that the cattle were fat and sleek before being moved to the market at San Bernardino."

By 1892 John Beatty was bringing stock in from his ranch near Weldon and had set up a slaughterhouse on a level area south of the present historic site of Garlock (Cow Wells). Nephews of Beatty who lived there during that period have identified the area where fence posts still stand, holding bits of the oldest type of barbed wire. In addition to running the butcher shop in Cow Wells, Beatty made fresh meat delivery once a week to way stations and prospectors along the route. When the gold rush started three and a half miles to the east, his business boomed, and deliveries increased to match the demands of the growing mining camps. To the meat business, Beatty added liquor, groceries, and mining supplies, housing the new merchandise in a tent near his one-room adobe home. The area was changing fast, people were moving in, and land was taking on value. The struggling merchants saw their efforts starting to pay a reasonable return.

Ore train out of one of the many tunnels of the Yellow Aster Mine.

(Courtesy Kern County Museum at Randsburg)

Ramsey Cox, working in Reed Gulch in Goler Canyon, discovered a $1,900 nugget and started the Goler gold rush of 1893. *(Author's collection)*

2

Nuggets on the Ground

With their imaginations fired by the glint of gold, few prospectors of the 1800s would have agreed with Charles Loring Brace, California traveler in 1869. In *The New West*, a guide to persons considering a move to California, Brace gave a less than glamorous picture of mining.

> If any of my readers have a lingering romance about a mining country, or the golden sands of California, they should travel through the southern mining counties. Mining at the best is a sort of devil's or ghoul's work on a landscape. The gay wild flowers of California are dug up as if with fresh-made graves; the rounded outline of hills is broken with heaps of dirt.
>
> What histories of tragic struggle with fortune and of defeat there are written in California! How many young men, for whom still fond hearts of sisters or mothers beat lovingly in vain, have fought the battle of life here unsuccessfully, and have died, as men know how to die, in solitude without a murmur or a groan. For in California, men who fail are too proud to return to the east or ask sympathy or help. I know nothing more touching than, in these scenes of former greed and insane money getting, to see occasionally a miner's grave. We can well imagine with what hope and courage that man had struck for fortune in this desolate place, where wealth sometimes comes in a day; how he tried and tried again, each time with lessening hope and weaker power; how the best years slipped away while he lost all that was worth having. I hope I may not be thought too melancholy in my picture of the mining country, but certainly, at this time, the California southern mining counties have a funeral aspect. Gold hunting in the ground, seems to curse not only the unhappy seekers, but nature itself.

No matter how harsh or disenchanting the written picture, gold fever, like other urges toward fortune and accomplishment, shut out the discomforts and dazzled the hopeful with promises of a better future.

The grip of gold on one traveler along the El Paso Mountains is part of the history of the area. Though various versions of the Goller (Goler) gold story conflict, a thread of similarity runs through them all. The local, most often repeated story is that John Goler was prospecting his way out of Death Valley in 1867, working south. In a small canyon of what is now eastern Kern County, he is supposed to have stopped at a spring for a drink. While kneeling at the spring, he was startled to see chunks of gold lying in the water below his lips.

Gathering a few nuggets as samples he hurried on to Los Angeles, where he hoped to interest someone in financing a prospecting trip. According to some versions of the story the fear of Indians caused him to hurry away, leaving his gun as a landmark. However, such behavior seems a bit strange in the light of his fear of Indians.

Goler is also said to have assured potential investors that he could go right to the spot where he had found the nuggets. A small map, in addition to the gold, was most convincing. His gun, left standing upright on a small hill out in the valley, would be a definite landmark, easy to spot.

A well-equipped party, led by Grant P. Cuddeback, was formed in Los Angeles, and Goler confidently led them northward into the Mojave Desert. Distrust began to grow when it became apparent that the map could fit any one of a hundred canyons, and finding a gun on a small raised area in an immense valley appeared impossible. The prospecting party covered a wide area, but found nothing; the disgruntled men threatened to hang Goler for misleading them.

Goler was so confident of finding the spring with the nuggets that he organized a second party, but the search brought no greater success. Legend has it that he kept hunting through the years, and early freighters thought they saw him and his burro prospecting along the eastern foothills of the El Paso Range. Jack Kurlitz, known as "Slate Range Jack," was quoted as having seen Goler many times in the vicinity of a canyon that later carried his name. Once when Slate Range Jack was camped at Mesquite Springs, Goler showed him gold found five miles to the east; this may have been the influencing factor in naming that canyon Goler when gold was discovered there and a rush started in 1893.

A far less romantic version of Goler's gold makes no mention of gun or map. William J. Manly in his *Jayhawker's Oath* wrote that John Goler and a German companion came through Death Valley in 1849. They reported that after leaving the Valley they found gold along the trail while searching for water. But Goler had been interested only in water and food. Moreover, when they reached Los Angeles, he started a wagon and blacksmith shop. After some time a party was organized with Goler as guide. But since he could not find any familiar landmarks, the party broke up and most of them returned to Los Angeles. Then a second group formed; they took a different route, but still found no gold.

Manly talked with Goler in 1862 and suggested that possibly the gold had been mica that had fooled many a prospector. Goler insisted that he would recognize gold because he had worked in a gold mine in Germany. Nevertheless, he still had no clues as to the location of the gold seen on the trip out of Death Valley. In 1895 Manly wrote about the rich specimens of the lost locations of the Gunsight Lead, Pegleg, Goler, and Alvord mines. Of the four, the Goler lost site was the place where enough nuggets to fill a pail could be gathered within an hour.

Newmark also wrote of Goler, describing him as a pioneer blacksmith and wagon maker in Los Angeles. His shop on Los Angeles Street was in a small adobe structure. Blacksmiths were scarce, and Goler's place was always busy; frequently pedestrians had to detour for some distance because of the number of vehicles awaiting repair and the horses waiting to be shod. By the late 1860s Goler-built wagons were being sold all over California, and his custom-built, one-horse-drawn wagon with an awning was a curiosity. In 1869 he invented the first successful velocipede, which caught the public fancy and grew into a craze.

In 1868 Goler and a friend organized the first political procession seen in Los Angeles and about that same time Los Angeles County gave him the contract to make their branding irons with the Letters *L.A.* But in 1871 a financial crash wiped out Goler's business along with many others.

Neill Wilson in *Silver Stampede* also mentioned John Goler as one of the '49 party who made it to Don Jose Salazar's ranch in the San Fernando Valley. It was at this ranch that Goler told of seeing the gold along the route. "For a quarter of a century the good Goler, who shod horses and made good substantial wagons and stage coaches for a living, repeated his story and between bouts at the bellows, spent his time and savings on expeditions that went off into the unrewarding desert."

While Goler searched for his lost fortune, others continued to find gold in amounts worth mining. An old letter dated 1887 tells of an expert stone mason, George Labb of San Joaquin Valley, who was brought into Cow Wells to erect an arrastra for crushing ore from the mines in the El Paso Mountains back of the settlement. The arrastra, a drag mill, had to be skillfully constructed or the gold would be lost. Stones, well-fitted and smooth, made the floor on which ore was crushed by huge granite boulders.

Cleanup at the end of one day or after many days was the term given to collecting the gold that had settled on the arrastra floor. The event was like the turn of the unpredictable roulette wheel; maybe one would hit it lucky and could pay bills, or maybe there would be little or nothing to compensate for the long hours of pick and shovel work before the ore even reached the arrastra.

When Ramsey Cox, working in Reed Gulch, an offshoot of Goler Canyon, discovered a nugget worth $1,900 and many smaller ones worth from $10 to $50 each, the stampede was on. News traveled fast. Overnight the whole canyon and its many-sided gullies were crawling with prospectors.

Anything that would serve as living quarters was quickly set up. The wide variety of shelters included dugouts; tents, combination rock, wood, and canvas shacks; wagon box roofs with tent sides; and blanket lean-tos tied to greasewood bushes and anchored with rocks.

On the heels of the miners came the girls of easy virtue. Mexican Nell from Tehachapi was the first to set up her tent. Soon the business establishments included Koehn's tent store to supply food and mining equipment for the dry gulchers and a saloon housed in a structure of wood sides and tent top. The bar was a rough plank resting on empty whiskey barrels.

Thus Goler Canyon, or Goler Gulch as it was frequently called, provided the area with a boom in 1893 that was the biggest thing to have hit the region up to that time. Here was a beckoning new field, and men poured in from other mining districts hoping that this would be their big chance. Charlie Koehn, a German-born engineer who first came to the area to install machinery at a gold mill in Mojave and stayed to develop Kane Springs way station, capitalized on the needs of the miners. He freighted supplies in and even began delivering mail at 25 cents an item.

The post office established at Koehn's headquarters, Kane Springs, in 1893 continued until it was absorbed by the Garlock office in 1896. Government records indicate that Koehn's earnings through the postal department for the first year of the Kane Springs post office were just over $50, but the total returns for this service to miners all over the area will never be known. Since he was not a man to continue an unprofitable business, it may be assumed that the charge per letter delivered made the service well worth the effort.

Though Koehn kept an eye on his bartender in Goler, he was not aware of the drama that was involving that handsome, mustached gentleman behind the bar. One of the girls of the night who had taken a special liking to the bartender courted his favor with home cooking. Each evening she brought him a hot, flavorsome dinner, and the gentleman began to grow fat and content. Although more girls were moving in, setting up their tents, and collecting a good portion of the miners' hard-earned gold, none appeared interested in the bartender. Therefore, no one noticed when he began to have a midnight dinner in addition to the earlier meal. But when at last the two cooks met, there was quite an uproar, and the bartender had no more home-cooked meals.

In Wallace Morgan's *History of Kern County* a news correspondent writing at Kane Springs is quoted as saying that by the spring of 1893 approximately $50,000 worth of gold had been taken out by a thousand or more men working Goler Canyon. He estimated that four-fifths of this gold had been found by less than a dozen men. The remaining one fifth was taken out by a small portion of the others. An unofficial census published at the close of 1893 listed 600 men, 14 women, 6 saloons, and 2 stores and estimated production to date worth between $125,000 and $200,000.

The freight route from the Searles borax mines to Mojave passed within a few rods of the Goler diggings. Stories that men deserted their teams and wagons to join the gold rush, though not verified, were probably true. The history of gold mining throughout the Southwest tells of men deserting everyone and everything to join the rush to riches. Once they tasted the excitement, few were strong enough to pull away from the mixture of gambling, poor liquor, gunplay, and the chance that the next day they would make their strike.

For all of their gold fever, however, the men of the Goler gold rush were far-sighted. They met and formed a mining district that in their enthusiasm took in a good portion of eastern Kern County. Minutes of the meeting at which the district was formed were sent to Bakersfield for official recording:

> Goler Mining Dist. Kern County, California. A meeting being called by the Miners of Coarse Gold Gulch in the county on the 15th day of March 1893, J. Waugerman acting as Chairman of the meeting and N. J. Ayer as Secretary.
>
> It was moved and carried that the name of Goler Mining District be the name of the district. Motion was carried by acclamation and the boundary lines were established commencing at Desert Station, about 23 miles north east of Mojave City and on the Owens River and Borax Road running 10 miles west from said Desert Station. Thence, north 20 miles, thence East 20 miles, thence South 20 miles, then West 20 miles to place of beginning. It was moved and carried that we be governed by the United States laws in regard to locating and monumenting claims and doing assessment work. It was moved and carried that the Recorder be allowed $1.00 for recording claims and all over 1 mile, 25¢ per mile mileage.
>
> Recorder's office situated in Coarse Gold Gulch, Goler Mining District, Kern County, California. By John Waugerman, Chairman, N. J. Ayer, Secretary. R. G. Willard was elected Recorder of said district.
>
> Names of all present:
>
> 1. J. Waugerman, Chairman
> 2. N. J. Ayer, Secretary
> 3. John S. Reed
> 4. D. Gilbert
> 5. P. W. Heap
> 6. C. L. Mecham
> 7. Henry Morse
> 8. G. F. Mecham
> 9. N. C. White
> 10. R. G. Willard

Although Ramsey Cox, discoverer of the big nugget that started the Goler rush, did not appear at the meeting, he stayed on, successfully mining the canyon long after those listed above had moved out to other developments.

With the influx of miners the Cow Wells settlement three miles away became the metropolis of the area. Stores and bars increased; tent and lean-tos of wood were scattered along the road, housing the families of men working in Goler. In 1894 Eugene Garlock of Tehachapi moved in his eight-stamp mill, taking advantage of the adequate supply of water for the operation of the mill.

Sage and other desert brush provided fuel to furnish steam power; the vegetation made a quick, hot fire with long-lasting coals. As ore brought to the mill increased, the name Cow Wells slid into the background as men talked of "the Garlock mill," "down to the Garlock," and then just "Garlock."

Conversation at night around campfires, at the corral, in the saloons, or at the Garlock was a mixture of freighting and prospecting talk. The life of a prospector seemed more desirable and easier than that of a freighter, but the uncertain income was a drawback. Neither life was for the weak, but the freighter was sure of pay at the end of the long, rough trip, while the prospector could go days without turning up enough gold to pay for beans.

Finley Buhn, who became one of the old

timers of the district, told of freighting to the strike in Goler and of mines opening up in the El Paso Range. It was his opinion that no real serious mining was done in Goler until after July 4, 1894, when a railroad strike turned a lot of unemployed men into prospectors.

Buhn made his first trip into Goler freighting lumber to the Van Brieson placer diggings. He stopped at Mesquite Springs to rest and water his teams and recalled that there were not only mesquite trees, but wild grapes and wild roses growing around the spring. In Goler were a number of cottonwood trees that were washed out later by a flash flood.

Buhn freighted with eight mules and horses, hauling two wagons with loads that averaged a ton per animal. He hauled hay, grain, lumber, and mining machinery at one cent per pound, and for a couple of years he did well. Other freighters moved in on the area and started cutting prices until at $8.00 a ton no one could make expenses. Buhn sold his outfit and went prospecting. In his words: "I made my share of good strikes but it all went over the tables, on the wheels, or on shoe polish. By shoe polish I mean the kind you get on the bottom of your shoes from resting on a brass rail."

Surface mining placer camps were going well by late 1894, and pay was in gold at the rate of $2.50 a day; usually gold was weighed out and given each man at the close of his day's work. From his leather poke a miner would pour gold out on the grocer's scales to pay for food, tobacco, or liquor. The saloons and gambling tables were eager takers, although not particularly accurate in their gold weighing. Men working their own claims put in all the daylight hours, while those working for others gave 10 hours for their pay, ate on their own time, and usually worked six or seven days a week.

Those who could afford it ate three meals a day at Mrs. Freeman's boarding house, a long narrow tar-paper-covered frame building. Both Mrs. Freeman and Mrs. Archie Martin moved their boarding houses out of Goler after a short time and continued their businesses in Garlock. The widow Freeman and her oldest son Bob managed to make a living for the family through her cooking and by 12-year-old Bob's using their team to haul water from Garlock to Randsburg for a dollar a 50-gallon barrel.

Before moving to Garlock young Freeman had picked up a few extra dollars by running errands; the best tips came from a woman living at the Solomon Nugent House, who ordered meals brought up from the boarding house. Freeman recalled that, "The older boys were always stopping me and asking what I heard or saw when I delivered the meal. Not until I was some years older did I understand their curiosity. Nugent ran a sporting house; the woman from Tehachapi was his favorite and was treated royally by having her meals delivered."

Mrs. Freeman's boarding house in Garlock, 1896. Bob, far right; mother at center. *(Courtesy Martin Engel)*

Like most boys, Freeman had an ideal, an older man whom he hoped to emulate in his own adulthood. That man was the swamper, Jim Wolf, of the colorful freighting wagons that daily traveled the valley. Freeman's greatest desire was to grow up and drive the sun-baked, dusty roads, handling the long teams, experiencing the dangers and glories of freighting. Claude Bohannon was another freighter whom he greatly admired, but the glamour dimmed as Freeman matured and Bohannon quit freighting to open a saloon in Randsburg.

Though Freeman was too young to take part in or be too concerned over quarrels, he did remark that there seemed always to be someone fighting over a mine claim or accusing another of "high-grading," taking the best ore off the top before the cleanup when all would share in the take as they had in the labor. "When there was real trouble, a meeting would be called and 30 or more would congregate to settle the matter peaceably."

Freeman said that for years he had gone around with a heavy secret, but, "Since everyone involved is dead, I'll tell something that happened while we were still up in Goler. Never through the years have I mentioned it to anyone. At night I often played checkers with the Stephen and James boys who lived across the draw from us. Sometimes they came over to my place; other times I went to one of theirs. On this particular night I was on my way to play with the Stephen boys. I'd had a lot of work to do around the boarding house so was late in getting started. On the way I saw men heisting ore out of a shaft between Koehn's store and Stephen's cookshack. I got off the path and worked my way to the Stephen's door; the boys drew me inside and hissed, 'You didn't see anything, you're playing cards!'

"I had recognized one of the most respected men in camp, but couldn't tell who was helping him; they could have been part of the tough bunch at the Black Hawk Mine. Everyone was afraid of them and no matter what happened they were the ones to get the blame. If they had been involved in this evening's high-grading and they found out I'd seen them, my mother's boarding house would be wrecked and we would have to flee for our lives. I didn't even admit to the Stephen boys that I'd seen anything.

"The next day Stephen and Koehn got into an argument, each accusing the other of heisting the high-grade ore that had been stored in long white sacks along the tunnel wall. Fifty sacks were missing and Koehn was frantic, asking everyone where they had been the night before and what they had seen or heard. He even asked me, a scared kid; but I denied seeing anything."

How much his mother knew of what was going on, young Freeman never knew, but soon after the high-grading incident she had her boarding house moved down to Garlock. It was a good move; business was better, and Bob's water hauling paid off. The only bad times were those when transients didn't pay their board bill after giving a hard-luck story and eating on credit for two or three weeks. Regular miners always paid weekly, and it was through them that Mrs. Freeman became familiar with good gold and the poor grade that was about half silver. She took her pay for board in gold and later became expert enough to buy gold.

In 1895 most of the miners in the placer diggings of eastern Kern County were like William Henderson. He, two brothers, and their father used a dry washer, the most popular method of getting gold at that time. Everything was done with manpower, and there were no milling costs to come out of whatever a day's work produced. It was a slow process for a man working alone; usually groups worked together, with each participant filling an important spot in the total process. Some worked underground following the gold deposit with pick and shovel; others brought the ore out to the surface by bucket or wheelbarrow, depending on the slant of the shaft. Outside the tunnel or shaft other men shoveled a steady stream of the freshly dug gravel and sand into a hopper that fed into dry washers operated by still more hands.

Henderson explained the importance of the dry washer operators: "The fellow running the dry washer had to know his business, keeping the washer level and adjusting the flow of gravel so that it traveled just fast enough to leave all the gold on the riffle board, and this changed with every new strata hit in the mine. A careless operator could let the gold flow by and out onto the dump; then the mine would be considered too poor to work."

At the end of day, "We were all present at cleanup time to see how the day's work had gone. You had to be careful of who you teamed up with or you could do all the work and the other fellow would high-grade or lift the big nuggets right off the riffle board all day, leaving very little to divide up that night." He added that high-grading was common and the most frequent cause of quarrels between partners. It was also commonly accepted that a man caught high-grading was asking to be shot.

The Henderson and Lackey men, both of whose

families lived in Garlock, worked an extensive placer digging north of the town site. Every one carried his share of the work, and they all made a fair living. The day young Henderson was working 150 feet back in a hole and came out with an 8-ounce nugget, the whole group had a good day's pay. Gold at that time was worth $20 an ounce, and the total was divided among seven workers.

Word of the nugget traveled to the other camps, and the following day a traveling photographer stopped by to record the scene and the miners involved. Those particular diggings were worked for three years more but never gave up another nugget that came near the 8-ounce size.

Placer diggings north of Garlock, 1896. William Henderson, far left, had dug out an 8-ounce nugget 150 feet back in the hole the day before this picture was taken. Others in picture: Jack Langdon on wheelbarrow; behind him, David Henderson with shovel at dry washer and Alex Henderson; unidentified man behind dry washer; Tom Lackey turning dry washer; Albert Lackey, center; and Abe Lackey with riffle board on the right.

(Author's collection)

3

Gold in Rand Mountain

It was 1895; prospectors and miners were working the El Paso Mountains, foothills, and canyons. Goler Canyon and the Summit Diggings on the east slope were the busiest areas, with men moving back and forth in the restless, endless search for the big strike.

The barren, unnamed mountains to the south and west had repulsed prospectors. They appeared to be practically unmineralized; to search for gold there would be a waste of a prospector's time. Considering all the wealth and activity that was in the future of those very mountains, it is difficult to realize that in 1895 the locality was not only nameless but virtually untouched.

Credit for the discovery of gold on what was to become Rand Mountain is usually given three men: F. M. Mooers, John Singleton, and Charles Burcham. In reality it was Mooers who had first seen signs of gold when he made a short prospecting trip there the previous year. In Mooers' own words, published by the *Brooklyn Eagle* in 1899, a newspaper that had employed Mooers in his youth:

> There were more than 2,000 people camped at Goler at one time, satisfied to make $2.00 to $5.00 a day in dry placer mining. The present town of Randsburg was only 10 mi. from the spot and yet there was not the slightest indication of the footprint of man or beast when I first visited that region which is now producing $200,000 in gold bullion per month. I was first there in 1894, made a hurried trip and had built a monument right in what proved to be the very heart of the bonanza.
>
> I had only a canteen of water with me at the time and was glad to get back to camp, with my tongue hanging out. Water is a very serious problem in this arid waste and one does not dare wander far from the source of supply.

It is currently reported that tenderfeet usually stumble into the best mineral finds. You can rest assured, however, that actual geological study found Randsburg and made it what it is. Many were the sleepless nights my partner (John Singleton) and myself spent in the study of where all the gold came from that was found in the placer mines.

The winter rains had been unusually severe and the gravel was in bed rock. Our drywashing machine had been standing idle for a month or more. The hungry coyote began to sniff in the tent door, too uncomfortably like the proverbial wolf. Our provisions man wore a sour look when he brought our monthly supply.

"I'm sorry, boys," he said rather despairingly.

"You'll have to dig up. Yet I don't see what you will dig. The ground won't dry for your machines until June. Now the fact is I did not come out here for my health and it takes money to buy groceries."

Mooers continued: "I remember passing a joke about its being a good omen, for you've got to strike bedrock before you reach gold, and we had struck bedrock in grub."

The "provisions man" was Charlie Koehn of Kane Springs, who was running a grocery wagon delivering service to dry-wash diggings outside the area served by his tent store at the mouth of Goler Canyon. Koehn took his pay in gold, weighed on a scale he carried on the wagon. He let miners run up bills and would help anyone in real need, but he also reminded them that his business could not stay in operation if the bills became too large.

In Mooers' article to the Eastern paper he explained at great length the geological theories and the conclusions that had led him to try prospecting the mountains to the south of Goler. Because they had no means of hauling water or equipment, he and Singleton had been forced to delay following up on their speculation that the desolate looking mountains could be the source of the gold found in the valley.

C. A. Burcham, the only person at the Summit Diggings with a team and wagon, was discouraged and making plans to go home to San Bernardino. Mooers and Singleton explained their ideas about the untouched area. At first Burcham was not receptive to their plan, but eventually he agreed to go as a partner with them on an exploratory trip. Burcham agreed to haul them and their equipment if they would provide the horse feed and three barrels of water. To quote Mooers:

We agreed at once and loaded up our picks, shovels, drills, powder, mortar and pestle, horn spoon to sample our rock, and with our scanty supply of rations started off.

To keep other prospectors from becoming interested in their movements, the three men let it be known that they were totally disgusted and were quitting the search for gold after one last look at their Goler claims. From Summit Diggings they drove northwest toward Goler Canyon. Well out of sight of the miners at Summit, Burcham turned the team south.

A primitive campsite was set up in a draw well up the side of the mountain (about where the Rand Community Dump is now located). Each man started out prospecting for the specks of gold that Mooers had seen the year before. Singleton knocked a chunk off a different looking rock and let out a yell as he stared at the amount of gold showing in the fresh break. Apparently, the colors Mooers had seen had weathered out of just such rock.

April 25, 1895, was the date. C. A. Burcham, F. M. Mooers, and John Singleton were the men. The gold discovered resembled that found in the Rand district of South Africa and thus the mountain became the Rand. This was a lode deposit. Unlike placer mining where individuals could work out the gold, this deposit would require considerable financing before it could develop into a paying operation.

Under ordinary circumstances such an apparently rich discovery would have called for a bit of celebrating, but at this point it had produced nothing to celebrate with and they were three broke prospectors. Past experience had taught them how fast news of gold could travel without the discoverers saying a word; so there was no time for savoring the future prospects.

Claims had to be staked and recorded as soon as possible; Mooers and Singleton were to remain on the spot to stake claims and get them on record while Burcham went to San Bernardino for supplies and financing. Burcham's wife, Dr. Rose

La Monte Burcham, had previously staked him to two years' prospecting with the understanding that half of what he found was to be hers, and it is probable that she was the financier to whom he now turned.

Upon reaching San Bernardino Burcham signed over to his wife one half of his one third share in the new discovery. Dr. Rose's enthusiasm equaled that of her husband and she cautioned against being too anxious to sell for what might seem like a good price. Mining history is full of the quick sales that left the discoverer with very little and the developer with a fortune. Dr. Rose did not want to see that happen. She would have been aghast if she had known what was taking place out on the desert at that very time.

During Burcham's absence two men from Bakersfield arrived at the Camp: O. B. Stanton, an investor, and Brown, an engineer from South Africa. Brown examined the ore and then suggested that Mooers and Singleton prospect higher on the mountain. To their delight the amount of gold increased as they moved upward.

Stanton pointed out that it would take a great deal of money to develop such a deposit and offered to erect a $10,000 mill for half interest in the mine and a six months' option to buy the property for $500,000. To the men who were down to their last cent and owed for the food they were eating, the offer was too good to turn down. They signed the agreement, assuring Stanton that Burcham would also be agreeable to signing for the remaining third.

They heard only the sound of big money, forgetting all the miners they had known who had sold out for too little. The sound of that big money was soon to turn to the sound of the courtroom with legal problems that threatened to destroy them. Both men would regret their rash move.

Burcham arrived back in camp on the Fourth of July and was greeted with the news that their discovery was already to pay off. Mooers and Singleton were momentarily disgusted with Burcham when he refused to sign and explained that half of his third share belonged to his wife. Burcham, however, apparently thought the deal was good because he immediately wrote Dr. Rose a letter to pave the way for selling, telling her that financiers were making offers. With a little development the mine would sell for at least $300,000. No mention was made of the paper O. B. Stanton held with the signatures of Mooers and Singleton.

Dr. Rose sensed the danger and decided that if she were in camp, some of the pressures from investors could be blocked. Losing no time, the lady doctor wrote her husband to meet her in Mojave, as close to the mine as she could get by rail. She allowed no time for a letter to return from Burcham, thus preventing any objections to her quitting the medical profession for the rough life of a mine camp. Packing a supply of medicines and sturdy clothing, Dr. Rose departed for the desert.

In later years she would laughingly tell of that trip. The train ride was slow, dusty, and tiring, but it was the 40 miles from Mojave to the camp that made the greatest impression. Burcham met the train with a team and light wagon; the July heat and dust relentlessly bore down on them between way stations where they stopped to rest the team and restore some of their own energy. Since it was dark when they reached Garlock, Burcham decided to camp for the night and to make the long climb up to the camp the following morning when the team would be refreshed.

With the blankets Burcham had brought along, they made camp near the wagon. Dr. Rose was so fatigued that even the hard ground felt comfortable and before long she was sound asleep.

"During the night I was awakened by someone nudging me, but it was too dark to see who; it wasn't Charles as I had to wake him up. He didn't seem the least concerned, explaining that it was probably the horses getting restless; he suggested that I throw them some hay and go to sleep."

Strange sounds of things moving about in the dark made it difficult for the young doctor to go back to sleep; when she remarked about it to her husband the next morning, he laughed and pointed to the hoof prints in the soft ground.

"We could have had our heads bashed in," the lady doctor gasped.

"Our heads are still on," her husband teased, but he was never able to convince her again that it was safe to sleep on the ground.

As they neared the new mine camp, Dr. Rose looked eagerly about, anxious to see what had been accomplished. Nothing Burcham had said had quite prepared her for the sight of two tents clinging to the mountainside beside a small, inconspicuous exploratory digging. But if the three discoverers hoped that the young woman would find the situation too rough and would depart for San Bernardino, they were due for a disappointment. Dr. Rose not only stayed; she immediately began to take an active part in camp life.

In the days that followed there could be no doubt in the minds of Burcham, Mooers, and Singleton that the one half of one third share holder was made of sturdy stuff and that she had a keen sense of business. Dr. Rose demonstrated the uncanny ability to see right through a shady deal, and many a sharp investor wished he had never met the doctor or felt the sting of her scorn.

Dr. Rose L. Burcham. *(Author's collection)*

She not only adjusted to tent life, cooking for the three partners, and washing in the limited amount of water, but also handled the financial records, keeping a sharp eye on expenditures. And although the men admired her business ability, they often smarted under her tight hold on funds, which apparently were from her savings.

Early development of the mine that was to become famous as the Yellow Aster was slow since there was no money for construction of large mills, wages for men, or buildings needed to house the overall operation. The first wagonload of ore was hauled down the steep mountainside to the Garlock mill where there was adequate water for processing. That mill run created one of the most often repeated stories of the early Rand Camp. Dr. Rose, wise to the ways of people and the use of money, admitted a complete lack of knowledge about the process of milling ore. Taking advantage of this lack, the three men came up with a plan for obtaining funds with which to celebrate their first run.

Singleton told the doctor that they needed a number of silver coins to put in the retort down at the Garlock mill as part of the process of separating the gold from the rock. Dr. Rose reluctantly handed over the coins, not wanting to delay the milling, but reminded him that their money was about depleted. With the money in their possession the three men couldn't wait to travel the 10 miles down to Garlock before starting to celebrate; so they went to the new tent saloon right in Rand Camp.

The silver coins were soon used up, and Singleton returned for more, saying that he understood that what they had was not nearly enough. The third request left no doubt in Dr. Rose's mind that the retort money was a ruse to cover the cost of drinks. It was reported that she gave the partners a real verbal tanning, but whenever the incident was mentioned in later years, the doctor would only laugh and add nothing to the story.

The first milling from the Yellow Aster Mine was keenly watched by the whole Rand area. A good run would mean jobs for many and for others it would hold out new hope that the claims they had staked in the same vicinity would be as rich. With everyone aware of the milling, the danger that the gold bullion would be hijacked while being hauled to Mojave for shipping was so great that the first shipment was delayed.

About this time Dr. Rose decided that she needed to do some shopping in Mojave and that she would accompany Mooers when he went to find out the procedure for disposing of the gold. The trip was made with Mooers driving the team while the dignified Dr. Rose jolted along on the spring wagon seat at his side, chatting about the scenery. No one guessed that concealed in the numerous folds of her long skirt the doctor held the heavy gold bricks that were to bring $850. She never intimated that she didn't trust Mooers with the transaction, but since he was prone to drink, it would seem that she made the trip not only to assure the safe delivery of the gold, but also the safe return of the money to the struggling young mine. Every penny was needed if the operation was to survive.

The first milling, in March 1896, processed 5 1/2 tons of ore, and the $850 paid costs plus a dividend of $100 to each owner to be used to defray expenses in Bakersfield while defending his property against O. B. Stanton.

As the legal hassle mounted, it was impossible to maintain ready cash available for a lengthy fight in court and still keep the mine in operation. Though Dr. Burcham, the one half of one third owner, had refused to agree to sharing the mine with outsiders, the time came when that seemed the only way to save even a portion of their gold mine much less keep the mine running. Under these circumstances, then, Pat Reddy, a smart and ruthless lawyer of Inyo County, managed to acquire an interest as payment for legal services.

In December of that year, 1896, Reddy took an option on the mine for $150,000. The option was to run 75 days; one half to be paid in February 1897, the balance the following month. No money was paid on either date, and when the option expired, Reddy tried to take possession by filing a damage suit of $50,000 against each of the owners. The lawyer thought that he had left the mine owners no escape except to turn the property over to him.

In May, two months after the option ran out, the original owners, now holding only thirteen twentieths, put the mine back into operation, leaving Reddy completely out of the management. By November, six months later, Reddy agreed to sell his rights back to the previous owners for $25,000 plus the nearly $10,000 in dividends due him for the portion he had owned for less than a year.

As soon as Reddy was out of the picture, incorporation was completed under the name of Yellow Aster Mining and Milling Company, with a capital of $1,000,000. Up to that time the operation had paid $30,000 in dividends.

Lawsuits plagued the young company from the very start. The paper Mooers and Singleton signed in Burcham's absence cost a small fortune to settle. Lawsuits over claim boundary disputes, equipment purchases that were not paid for when due, and unpaid wages all dipped into the seeming wealth. The owners, but most often Dr. Rose herself, made trips to Los Angeles and San Francisco to settle claims or represent the company. She often moved in on labor disputes when the men became too angered to work out the problem.

The four people most concerned in the development of the Yellow Aster and its millions appeared to get along well together, at least as far as the public saw, though they were complete opposites in personality and background.

Mooers' life sounds like a Horatio Alger adventure: the young hero suffers untold hardships but through resolute effort wins fame and fortune. In 1860 young Mooers had worked for the Demas Barnes Drug Company in New York, then moved to a position as head bookkeeper in the business department of the *Brooklyn Eagle*. It was a good position and the employment security was ideal, but Mooers craved action.

He left the *Brooklyn Eagle* and went west toward the adventure he dreamed about. Friends seldom heard from him as he moved about continuously, pausing in one place only long enough to work a while at the local newspaper and rebuild his finances so that he could move on. His wanderings took him from Chicago, west to

Montana and Wyoming, then south into New Mexico and Arizona. Newspaper reporting or any other work around a paper was his major means of making a living, but his true love was prospecting the land he passed through.

Mooers' own account to the *Brooklyn Eagle* told of the rough life in the mining camps and the many times he had been near starvation while in search of gold. Eventually he tried the California prospects and was in Tehachapi editing the paper there when he and Singleton, a millwright also working in Tehachapi, teamed up to try their luck in Goler Canyon.

John Singleton, born in 1847 in Sunny, Tennessee, went to work for an uncle in Texas after he finished common school. In his position in the brokerage office of his uncle's cotton firm young Singleton learned business methods. In the fall of 1869 when the Union Pacific Railroad was completed and there was big mining excitement at White Pine, Nevada, Singleton moved west to seek his fortune in gold. Instead of stopping in Nevada, he continued on to California and made and lost considerable sums in a number of camps by always trying to develop other mines that promised to give bigger returns. During the lean periods he worked as carpenter and millwright in the mining areas.

The third man involved in the discovery of the Yellow Aster Mine, Charles Austin Burcham, was a native Californian, born at Vallejo, November 6, 1859. In 1880 when he was 21, Burcham, his father, and an older brother purchased a 4,000-acre stock ranch 25 miles from San Bernardino. For six years the three of them worked 2,000 cattle, improved the buildings, and developed the project into a paying business. For the next four years young Burcham ran the ranch by himself and opened a wholesale/retail provisions business in San Bernardino.

In 1887 Burcham disposed of some of his cattle and property. Water rights were granted to Major Bonebrake for use in developing Hesperia. In December of that same year Burcham married Dr. Rose La Monte in Los Angeles. Dr. Rose was a recognized physician in San Bernardino and active in social functions of that community and Los Angeles.

Burcham's first interest in mining came during the Comstock boom in Nevada; he became one of the many investors who hoped to join those who were becoming millionaires overnight. The gold fever would give him no peace of mind until in 1894 he sold his cattle and other ranching interests to devote all his time to mining. Dr. Rose agreed to grubstake him for two years with the understanding that if he had struck nothing at the end of that period, he would return to San Bernardino and forget mining.

During his lifetime Burcham was not only associated with the Yellow Aster Mine in Randsburg, but was also President of the Calico Mining Company, President of the Rand Mercantile Company, President of the Phoenix Development Company, and an investor in mining projects all over California and Nevada.

Burcham's wife, Dr. Rose La Monte Burcham, one half of one third owner of the Yellow Aster Mine was the daughter of a Rochester, New York, physician. Against her father's advice she was determined to acquire a scientific education. He gave her access to his library but no other encouragement. Teaching school to earn enough for expenses, she attended Rochester Free Academy and received a diploma in 1882, went on to get a medical education at the Eclectic Medical Institute of Cincinnati, Ohio, and graduated in 1884. After one year as intern in the Cincinnati Hospital she decided to go west where "even a woman doctor" was acceptable.

Dr. Rose was the only woman appearing in the pages of *Men of Achievement in the Great Southwest*, published by the *Los Angeles Times* in 1904. This recognition was based upon her being the only successful woman mining operator in the southwest, "having demonstrated that she was endowed with executive ability of a high order."

C. A. Burcham, John Singleton, and F. M. Mooers, discoverers of the Yellow Aster Mine.

(Author's collection)

Butte Avenue, Randsburg, before the big fire in 1898.
(Los Angeles County Museum collection)

4

Rand Camp

It wasn't long after Burcham, Mooers, and Singleton started developing their gold claims at Rand Camp that other prospectors moved in, first to see what was going on and then to set up their own tents and stake claims along the hillsides and down the washes. Close on the heels of the prospectors and miners came the saloons and the girls. Even before the new mining venture had proved its value, there were already tent businesses to give the miner recreation and take his money. An early resident remarked, "It seemed as if every other business that started up in the early days of Rand Camp was a saloon and most of them had a gambling table."

Here were the usual ingredients of a mine camp. Men continued to flow in, too busy searching for a fortune to care or even think about how the town should build. Streets just happened where it was most convenient, where wagons had already cut a path, or where buildings happened to cluster to shelter each other from the wind. Tents and eventually wood structures followed the curve of wagon ruts, and thus many stood at odd angles to each other. For convenience and companionship, shelters were pushed close together, and later, store buildings were built so near to each other that they appeared to be attached. In spite of a whole mountain and miles of desert, the camp huddled in one small draw that led up to the mine diggings.

As families started arriving, the camp began to take on a settled look. Even in its infancy Rand Camp had status divisions. An invisible line divided the so-called cultured section from the miners' tents, dugouts of Fiddlers Gulch, the saloons, and the red-light district. A small segment of the growing settlement soon started a movement for churches, a school, and events or entertainment that would compete with the saloons, gambling, and girls.

On December 20, 1895, eight months after the discovery of the Yellow Aster Mine, Rand Camp was officially organized by 33 residents. A citizens committee was formed to study such community problems as the need for law enforcement and cooperative action for the good of all.

Nevertheless, life in Rand Camp was rough that first winter. Old-timers told of living in tents and camping under wagons and of trying to burn wet sage to boil coffee while a tarp hung down from the wagon bed to keep the blowing snow from putting out the struggling fire. Many had arrived thinking that the desert is always hot and had failed to bring adequate clothing or bedding.

Wagons creaked and groaned through the mud or dust of the street. Bells rang like miniature chimes as the lead horses moved, and chains clanged with a rhythm all their own. At night the sounds changed completely. Melody from the tinny piano in the Elite Dance Hall mingled with that of bow and strings up in Fiddlers Gulch. Loud voices, raw-liquored laughter and an occasional gunshot made the camp a lively place after sundown. All work and no play was not the code of this mine camp. Present-day strangers who ask old-timers how they stood the dull life before movies and TV get a stock answer: "Why, we had fun! We lived it."

Clyde Kuffel, an early-day freighter of the area and one of Randsburg's first residents, used to explain that to most miners the brand of liquor made little difference; it was the sound of voices in a saloon that gave you a lift after being alone, down in a mine shaft all day. The friendship of the bartender, the laughter of the girls, and story swapping with other miners all added up to a pleasant evening and made the day of hard work and uncomfortable living more bearable.

One night Kuffel came into Rand Camp after a cleanup of a good gold pocket he had been working for some time. Thirsty for refreshment and companionship he headed for the Stein Beer Parlor and placed his bar of gold where all could admire it while he had a few drinks. When an argument started later in the evening, Kuffel threw his gold brick at the offender. Either he was a poor pitcher or the drinks upset his aim, for the gold sailed right out through a newly installed window, the first to be found in a Rand Camp business establishment.

Windows were a luxury, hard to get and expensive by the time they had been freighted by team and wagon from Los Angeles. It took most of Kuffel's gold brick to pay for replacement of the window. A short time later when he became owner of that particular emporium, he called it Kuffel's Gold Brick Saloon.

The moments of pleasure were not for adults alone; the kids had their fun, too. Almost anyone could catch a burro around camp and go for a ride. The boys thought it great fun to try their hand at dynamite; and with all the blasting going on at the mines, no one knew the difference.

Celebrations were enjoyed by adults and children alike. Rand Camp's first Fourth of July celebration in 1896 was a real inexpensive blast. Clyde Kuffel offered to get rid of some aged dynamite as an opening salute to the day. The deteriorated explosive had reached an extremely dangerous stage and the slightest jolt could set it off.

Keyed up in anticipation of the exciting and dangerous salute to the historic day everyone arose at daybreak to witness the event. No one breathed easily as Kuffel's wagon was loaded and went jolting down the street with the dynamite shifting as the wagon wheels found every bump in the rutted path. Kuffel had experienced many dangers in his years of freighting, but people felt that this time he had overextended his luck.

Finally the wagon rounded the hill and went out of sight. The camp was safe, but chances were no one would see Kuffel in one piece again. How he managed to unload the stuff and remove his team a safe distance before igniting it has remained a mystery. The blast shook the whole mountain, and the few windows in the camp were shattered. The town waited; then some of the men mounted their horses to ride out to see what had happened. Just then Kuffel drove into sight.

That first Fourth of July observance and many to follow became the big event of the year. Through organization the timing of events improved, but generally the programs were similar year after year. Local musicians formed the band at the head of a parade of decorated wagons, bicycles, and burros. A typical program read:

1776 - 1900

GLORIOUS 4th OF JULY

JOHANNESBURG

CALIFORNIA

Everybody Welcome

Opening Salute at daylight by the battleship Olympia Dewey and Wife in command.

GRAND PARADE to RANDSBURG

The $5,000 assortment of Fireworks is to be in charge of a Special Committee of 100, under the direct supervision of

The Great and Only SAM BEERS the World-Renowned Fire Wizard,

Imported at an Enormous Expense.—Fire Dep[illegible]t in attendance.

The **News of the Day** will be received at the **Randsburg R.R. Depot** by the Lightning Operator, **M. R. Lopez,** who will also receive the first telegraphic message from Mars at 9 G.M.

The Execution of 10,000 Chinese

Will take place in front of the **Johannesburg Hotel** from 9 G.M. to 5 G.E. Their heads to be served at the Hotel, a la mode. No extra charge for this special dish. These Chinese were imported direct from Tien Sien for this occasion.

General Buller, commanding two British regiments, will dress parade during the execution of said Chinese.

The Bonfire Exhibit will be given by the representative of the **KERN COUNTY LAND CO.,** who will burn 100 Tons of Hay in front of their Warehouse on Broadway.

The Sham Battle will be a Bombardment between the Battleships Red Dog and Pinmore, on their respective tailing ponds, Griffith and Godsmark in command.

The Foot Race will be given by Mining Experts, the admission fee being a pair of leggings; the prize, the Yellow Aster Mine and Mill.

The Declaration of Intentions will be read by the Secretary of the Combined Water Works, who will come from Los Angeles for that purpose, where he is now enjoying the cooling zephyrs from mother ocean.

A Great Exhibition of Strength will be given by the **Miller Brothers,** who will carry the Johannesburg Hotel to Randsburg and back, without stop or water. W. S. Miller incidentally consuming a barrel of his world-famed whiskey en route.

S. H. Fairchilds will exhibit his Trick Mules at any old time when requested.

Teagle Brothers will present a Car Load of Groceries and a Bunch of Segars to the person holding the lucky number.

J. N. Carrow will Shoe a Ten-Horse Team in Ten Minutes by the Town Clock.

Capt. Colson will Cut 'Steen Thousand Feet of Lumber against time, before breakfast, to enable him to thoroughly enjoy his supper.

The Randsburg R.R. Co. will run Three Special Trains, to take care of the crowds expected from the Suburbs.

Orator of the Day and Deputy Constable - - - - W. W. Godsmark
Marshal and Aids - - - - - - - A. Benham et al

Citizens in Uniform, Officers on Foot, Ladies on Horseback and Babies in Automobiles.

For children the day was full of excitement, from taking part in the parade or running along shouting and laughing at the participants, to taking part in the Exercises. The contests always started off with the activities for fat men and women coming first before it got too hot. If you really wanted the prize money, you did a bit of practicing before the day arrived; Clementine Lovett wanted an Indian dress, but there was no family money for such a luxury so she got up early morning after morning to race up and down the mountainside. When the Fourth arrived she was ready. She won the girls' race that year, and the Indian dress was hers.

By nightfall most youngsters were happy to sit down to watch the fireworks, going to sleep before it was over. Then, half awake and half guided they managed to reach the family wagon to sleep through the dance or the long rough ride home. If the adults stayed for the dress-up dance, the children slept in the wagon boxes, on buggy seats, or at a friend's home while their parents danced till daybreak. It would be six months before most of them would get together for another celebration.

The New Year's get-together was the next big event. William C. Wilkerson, who arrived in Rand Camp as a boy of 16 told of the noisy farewell to 1896 and the roaring welcome for 1897. "Christmas was for the family. Some of the men went on working in their mines but stopped early for a special dinner. Some visited neighbors or had relatives from distant communities, but on the whole it was a quiet observance. But New Year's was different; they really blew the lid off!"

People came into Rand Camp from miles around to attend the New Year's dinner and dance. It was a happy time with people greeting and visiting with those they hadn't seen for months. Some brought dynamite to liven up the scene, and all evening the explosions shook the place as lighted sticks were thrown into the air. Between explosions, guns were shot off just for the joy of it. Everyone shouted to be heard, competing with the clatter of arriving vehicles, shouting children, barking dogs, explosions, and music floating out of the saloons.

By the time 1897 arrived, fatigue had caught up with the children. They slept curled up under blankets in the wagon or leaning against a parent at the dance. Liquor had built tempers to a high pitch, and men were spoiling for a fight. But in spite of threatened shootings, by the following morning it was evident that no one had been seriously injured and a few black eyes were the only visible results of the fist fights.

Whether they celebrated so hard as justified relaxation from 10-hour work days, seven days a week, or because the coming year held out more promise, the spirit of the evening was enjoyed by all. 1896 had started with Rand mine no bigger than a 9- by 12-foot room, a small settlement of tents, dugouts, three canvas and frame saloons, a tent assay office, and a tent boarding house. The only all-wood structure in the camp was a building moved up from Cow Wells and converted into a store. Water was hauled up from the same settlement in wooden barrels; greasewood roots washed out in flash floods served as firewood. During that summer, water, beer, and food were cooled in an evaporative setup of burlap sacks over a frame or apple box with water dripping over it.

By the time 1897 arrived the boarding house had moved into a newly erected wooden building and the dance hall had a real wood floor and walls. The population was on the increase and new businesses were starting to move in to supply the growing demand.

A first-hand view of the new year was given by Ira Gordon, who was 19 years old when he and his brother arrived in Rand Camp to operate a livery stable. Things were really booming early in 1897, and their stable of 35 horses and a number of burros was a busy place. A team and buggy rented for $10 a day for long trips; saddle horses cost $3 to $5 a day, depending on where they were to be ridden. A burro brought 50¢ to $1 a day, depending on use. To rent a burro for prospecting was cheaper than to rent one for transporting ore.

For investors or mining engineers who arrived in town and wanted a team and buggy just to get to the mines, the charge was usually only $5. Sometimes the individual hiring the rig would drive, but since these visitors didn't know their way around, they often hired Ira as driver and guide, thus making the business more lucrative. Frequently his passengers were people who were or were becoming famous in the mining world.

The stable was on the corner opposite the present post office and ran through to the street below. Across a vacant lot from the stable was Big Ella's where six to eight good looking girls plied their trade. The Wells Fargo stage stop was at the side of the stable and the first lodging house for travelers was on up the street about where the present Kern County Museum stands; the theater was also in that area.

One of the most interesting transactions Gordon made before he and his brother sold out in 1900, was with B. Bee, a mining engineer for Wells Fargo. Bee rented a span of mules and a wagon to go to Ballarat. He expected to return in a week, but day after day passed with no sign of the engineer. Gordon and his brother scratched off the team and wagon as loss, figuring that the man had driven off the road in a sandstorm or had tried to drive over unfamiliar roads at night and that he and the mules were dead. Exactly 125 days after hiring the rig, the engineer pulled the team to a stop in front of the stable, threw the lines to Ira, and went inside to settle his bill. Paying the regular $10 a day, he left without explaining why he had not sent word to the Gordons that he planned to keep the outfit for four months. The animals were in top condition; so it was obvious that it was not a case of getting lost.

"There was also an incident that was quickly hushed," Gordon said as he talked of the Rand Camp of 1897. "A fellow by the name of Scott drove for Mooers and Singleton of the Yellow Aster. When they wanted to go to their other mining interests in the area or even up to the Yellow Aster, the team was supposed to be ready and waiting. No one knew what the trouble was, but Mooers and Scott were continually quarreling. One afternoon they both happened to be in the Bit of Heaven saloon, standing at opposite ends of the bar. Remarks flew back and forth until Mooers became so angered that he rushed at Scott, knocking him down. In falling, Scott's head hit the new brass rail and the blow killed him. The notice of death read: Death due to accidental fall."

Traffic into the area was so heavy that four or five stages from Mojave via Garlock to Randsburg made the 8-hour trip daily. Hotel boarding house accommodations had to be increased but at times were not equal to the demand and drummers would hire a rig to take them back to Mojave when they had completed their rounds.

Operating the stable was only part of the activity the Gordon brothers became involved in. At one time they grubstaked a prospector who was sure there was gold over in the mountains back of Searles borax works. "Before we realized what was going on, the fellow was back in town drinking, boasting, and tossing gold around. We got him away from his eager audience and reminded him of our share for the grubstake; before he could spill any more about the location, we bought him out. When the fellow sobered up, he took us to the canyon this side of the Dean Mine."

The Gordon mine had rich ore at the start. Twenty tons was freighted to Johannesburg and shipped to the Selby Smelters. "After all expenses were paid, we cleared $2,800, but turned right around and spent it all on shafts, drifts, and a stamp mill. At 42 feet the gold pinched out until there wasn't enough ore to mill. Like a multitude of others, we had poured every cent we got right back into the ground and we went broke."

One of young Gordon's most unusual experiences was associated with a druggist working in the Rand Drug Store, the building that became known in recent years as Rinaldi Market, the ruins of which still stand at the entrance to Randsburg's main street. The wife of one of the two druggists passed away, and she was to be buried in Los Angeles. With summer heat at its peak, transportation of the body posed many problems. The plan was for the two druggists to go on ahead, stopping halfway to Mojave to rest the team. The body, packed in half-melted ice that had been hauled in from Mojave, would leave Randsburg later with Ira Gordon as driver of the light wagon. At the half way point Gordon would pick up the rested team and the husband of the deceased. The other druggist would return to Randsburg as soon as Ira's team had rested. In this manner they hoped to catch the train out of Mojave with the least delay en route from Randsburg.

After leaving the halfway point in the intense heat, Gordon pushed the team as fast as possible because the ice was melting fast and running in a stream out of the wagon. Six miles from the Borax Company barns one of the horses died, and the druggist insisted that he should be the one to ride into Mojave to get help. Young Gordon thought otherwise, jumped on the remaining animal, and rode off to the cries of the man, "Don't leave me here alone with the dead!"

Gordon went to the Borax Company corral, caught a mule, and led it back to the wagon. By the time they reached Mojave, the train was gone, so Gordon went to Sheriff Underhill for help. The officer sent out a call for women to help repack the body in ice. When the corral manager reported a theft, the sheriff didn't admit that he knew anything about it, but suggested that the fellow take better care of his job, unless it was a case of his not being able to count. Later, he told what had really happened and pointed across the street where the mule was enjoying his fill in a feed lot.

Like the Gordons, Joe Short arrived in the area in 1896, but settled down in the valley at Garlock. A former undersheriff in Bakersfield, Short hoped gold mining would provide the needed income for the six of his eight children still living at home. Thurman, one of the boys who went to school in Garlock, still visits the area and recalls how everyone in the family had to work to make ends meet. He and a brother would go to the placer field with their father, and while he dry-washed promising looking surface diggings, the boys would go down in the honeycomb maze of old mine workings of the early Goler gold rush. Very carefully they would remove the gold-bearing streak left in the support pillars, leaving only a narrow strip, sometimes no thicker than their thumb, to keep the support from crumbling down on them. They carried their diggings out in lard buckets for their father to run through the dry washer.

Thurman added that, "Between what our father worked out of his diggings and what we brought out of the old mines we got enough gold to keep us eating. As a matter of fact we lived pretty well for that time; we even had a tar paper house when most of our neighbors lived in tents or dugouts."

One of Thurman's fondest memories revolves around an old miner named Vedder, who was an outstanding violinist. The whole town enjoyed his music, but Vedder was not always in a mood to play. The boys found out that he talked to spirits and that if the spirits wanted music, he would play the violin. Presented with this challenge the boys decided that the townspeople of Garlock were just as deserving of good music as were the spirits. After a few experiments they discovered that the whine of a rosined string fastened to Vedder's window had the right effect, and violin music soon floated out on the desert night.

In 1896 Garlock was having its boom, too. The gold mill erected by Eugene Garlock was busy, with wagons of ore waiting their turn. Other mills were added; the Kelly, Smith, Henry, McKernan, and Visalia were the large ones, but small operations were springing up all over the area. The process of milling was slow, and recovery was not too good; so some miners thought they would benefit by doing their own milling, even on a small scale. Stories of theft by the mill were rampant; whether they were true or not, the hard-working miner was concerned over every ounce the mill produced.

Except for the Garlock mills the only sizeable operation at the time was Charlie Koehn's at Kane Springs, where he, too, had adequate water for the process. Koehn also owned a freighting string that hauled the ore down to his mill; so he made money from the time the ore left the mine. It was reported that he milled $440,000 worth of bullion in two years. Though there seem to be no figures on the amount of gold milled at the six Garlock operations, a rough estimate can be reached considering the fact that they were similar or larger than Koehn's.

Six mills running full time meant employment and more families moving to Garlock. Jim McGinn opened a store in addition to the two already doing business. Dr. W. H. Wright arrived and built a wood structure that housed his drug store and doctor's office and later also served as post office;

a lean-to sufficed as living quarters. In addition to being doctor he was community counselor, preacher, prospector, and mining investment advisor. A. J. Doty of Mountain Home in sequoia country built a two-story hotel on the main street. Ed Maginnis, agent for Wells Fargo, had been handling the Goler district gold through his office and now had considerably more to ship.

As a boy, James Nosser, Jr., would watch his father's wagon slide down the steep grade just out of Randsburg en route to the Garlock mill. He was not invited to ride along, and within the family circle the town of Garlock was never mentioned. His father's sister Sarah had married John Miller, who in 1896 built a pretentious rock building in Garlock to house a bar, a gambling room, and tiny rooms for the girls. The exterior of the building looked like a bank, and in a way Miller was a banker. He bought the miners' gold in his office at the back of the building and then collected what he had paid out as the miner worked his way past the doors of the girls' rooms, past the gambling tables, and past the bar before reaching the outside with what little was left of his gold. Miller's business was well located just across the street from the stage stop at Doty's Hotel, only a few steps away for the thirsty traveler. For some this was the end of the trip; for others there was still the 10-mile climb up to Rand Camp. Refreshments at Miller's helped ease the last weary miles.

Elite Theatre, Randsburg, 1898, had wood front and canvas sides. *(Los Angeles County Museum collection)*

5

People and Progress

A timid young woman, Kate Pfadenhover came to Rand Camp seeking employment in what she understood was a fast-growing community needing women workers. She stepped off the stage to an immediate inquiry: Would she be interested in work in a restaurant? Without waiting for a reply, the restaurant owner added the inducement of a room in a boarding house. Delighted to have employment so soon, Kate accepted the job right on the spot.

As she waited tables she noticed how busy everyone seemed, the number of wagons and buggies filling the street, and the happy sound of shouting and laughter as barefoot boys dodged in and out among the moving teams. This was going to be a nice place to live and work.

That night Kate saw the other life of the mining camp. The laughter now came from the girls of the night, and the shouting was in the deep tones of angry men. Piano music came from most of the saloon doors and mingled with loud talk, barking dogs, an occasional shot, and the cheers of bystanders as a street fight developed. Kate was so frightened at the wild life in the street outside her window that she moved the bigger pieces of furniture against the door and window.

The restaurant work and the pay were fine, but at the end of each day Kate would rush to her room and barricade the door in fear for her life. The day the merchants and the restaurant owner closed the metal shutters over the windows against a shooting scrape in the street, the young woman decided that no amount of pay could compensate for living in such a dangerous place. She packed up and caught the next stage out.

Riding the stage was an experience in itself.

Drawn by two, four, or six horses the stage bounced passengers skyward when the wheels hit ruts or rocks; the remainder of the time the swinging, swaying motion was enough to make some passengers "seasick." The larger stages carried as many as nine passengers inside, with two to four on top when necessary. Luggage and mail were fastened on the back, and the most valuable articles such as gold or payrolls were tucked under the driver's feet. Horses were pushed, traveling as fast as the rutted road or sandy areas would permit. Fresh horses were picked up at way stations while passengers stretched their cramped, weary bodies. Some travelers carried baskets or collapsible suitcases filled with food; others had to rely on what the way stations had to offer. The usual menu featured tough beef, boiled potatoes, stewed beans, and a dish made of dried fruit. The meal was served on a plank table covered with oilcloth; rough planks served as benches; and at night a coal-oil lamp sent a yellow flickering light over the scene.

Tom Garrity, who ran a restaurant in Randsburg for about 40 years, came to the mine camp the summer of 1897 and described the stage trip: "Mojave was the jumping off place for the diggings. We pulled out at 3 a.m. in a big Concord stage with a team of four. All the rest of the passengers were boomers loaded with whiskey, and drinks were passed around every time we hit a sharp curve. The driver refused to drink.

"About noon we pulled into Randsburg and every one got out. When the driver started to turn around, he cut it a bit short because of the narrow street and the stage tipped over. A whiskey drummer called Gibbons yelled, 'I'll bet you the drinks you can't do that again.'

" 'And I'll go you,' the driver shouted back as the watching men helped him pull the stage upright. The driver drove the stage down the street and then came back and flipped it over again. The crowd roared, and every one went into the saloon. The drummer stepped to the bar, and every one in the house had a drink at his expense."

Garrity described Randsburg at the time of his arrival as a sprawling settlement of rutted streets, several stores and restaurants, lots of saloons; most of the miners lived in tents and only a few families had wood shells for houses. "Anything you lived in was hot in summer and cold in winter. The Yellow Aster and a number of small mines were producing, but there was little chance to stake a claim as everything for miles around was already taken up."

In spite of the publicity in the Los Angeles papers that had encouraged Garrity to make the trip, there was little chance of discovering or becoming the owner of a mine. "It looked as if they needed restaurants more than anything else; so I started one. In my 42 years here I've owned seven restaurants, sold a half interest in two, given one away, and lost four in the big fires."

In 1897 there was a lot of cash floating around the area. Life was geared to spending and hoping that the big fortune would appear the next day. Miners coming in from other boom towns remarked that on the whole Rand was more law-abiding than many camps they had been in. Rough fun was accepted, but when a character got too far out of line, some of the town fathers or members of the Citizens Committee would escort the fellow to the edge of town and give him the choice of a bottle of water and the road to Mojave, or the road over the hill to the cemetery. When anybody asked about a person who wasn't around anymore, the common expression was: "He took a bottle of water."

Publicity in the Los Angeles and San Francisco papers was credited with drawing the undesirable as well as the serious miner. The population reached 1,500, with 40 to 50 people arriving every day. The stages, for a fare of $3.75 from Mojave, never had an empty seat. This is the kind of promotion material that started the population boom of '97:

> Two stage companies run from two to four coaches each daily from both points, Kramer and Mojave as required by the traveling public. As in all mining camps the hastily built shanties and tents are rapidly being replaced by more substantial buildings. Hotel accommodations are good, there being two first class hotels, the St. Elmo and the Occidental, while good comfortable lodging houses are in abundance, the most popular being the Rand and Russ.
>
> The first mines discovered were the Rand, Olympus, Trilby and Yellow Aster, now called the

Rand group and are situated about a 1/2 mile southwest of town. These mines are now bonded for $200,000. The other principal mines are the Butte, Wedge, Kenyon, Little Butte, J.I.C., Monkey Wrench, and Excelsior, all located on the edge of town, while the King Solomon group, Val Verde, Jolly Girls, Alameda, Ruby and Sheldon are about 1 1/2 miles due east. The King Solomon is producing good milling ore at a depth of 160 ft, while the Val Verde is down over 100 feet and milling ore paying from $20 to $50 per ton. The Wedge is down 200 ft and within the last 30 days struck a pay chute assaying $1440 per ton. The Kenyon is down 150 ft and very recently took out ore assaying well over $1000 per ton. Both are now working in good paying ore and are confident of tapping a big body of ore before they get much deeper. The Butte, which has been one of the largest producers is down 160 ft and recently let a contract for an additional 100 ft more.

The St. Elmo is located about 6 miles south of Randsburg; is capitalized at $2,500,000 and sinking 3 shafts, one of which is down 140 ft, while a double compartment shaft is down nearly 100 ft. A ten stamp mill and cyanide plant will be in place on this property at an early date as the preparatory work is now being done for such improvements. At present there is sufficient labor in the district but there is a great field open to investors who have a few hundred or a few thousand dollars, as a great many men have from one to ten locations and are without means to do developing and are willing to give part of their holdings to persons furnishing money for development. The country is highly mineralized in all directions for 10 miles from town. The season for drywashing is now at hand and much gold is being taken out by that process. There can be no question as to the future prosperity of the Rand Mining Dist.

Respectfully submitted,

Thomas & Rogels, St. Elmo Hotel
Pratt & Pearson, Occidental Hotel
Mrs. M.A. Johnson, Rand House
Mrs. T.M. Harrington, Russ House
Greive & Gordon, Livery Stable
M.S. Lane, Manager Crystal Ice Co.
A.J. Flood, Chemist & Assayer
U.T. Clotfelder, Lawyer
Geo. Pridham, Agent, Wells Fargo & Co.
Tootacre & Wegman, Lumbermen
F.W. Rundel, Harness Maker
F.R. Seibert, M.D.
Palace Billiard Parlor & Bar, W.A. Coon & Co.
W.H. Hiveren, Mgr. San Joaquin Valley Supply Co.
F.W. Brendel, Wines & Liquor, 'National'
Price & Hopper, White Fawn Saloon
George W. McPherson & Co. Realestate & Mining Brokers
B.M. Birdsall, Gen. Merchandise
Hammond & Co. Gen. Merchandise
G. McK. Bevan, Agent, Maier & Zobelein
J.W. Ragesdale, Mining Broker
C. Linkenbach, Lawyer
Emerson Stron, Lumberman
Norton O'Bear, Grocer
F.C. Hempstead, Druggist
W.C. Wilson, Gen. Merchandise
Co. R.F. Haddord, Wines & Liquors
Callaghan & Endl, Wines & Liquors
Adams & Bailey, Capitol Saloon
Orpheus Theatre
John Crawford, Wines & Liquor

Risk capital for mining investment was not difficult to obtain. Big business on Spring Street in Los Angeles, the Wall Street of the West, was interested and promoting. The names Yellow Aster, Pinmore, Kenyon, Big Butte, King Solomon, Baltic, Buckboard, and Sunshine were becoming familiar to everyone. Men on the street talked with authority about the great wealth of the district that would be unearthed to the advantage of anyone in on the development. Little or no mention was made of the myriad of holes, tunnels, and shafts that did not produce a cent for the hours of drilling, blasting, and plain hard work with pick and shovel.

When a rich pocket showed up, the mine owners and miners all benefited and the public was quickly aware of the good strike. High-grading was common among the working men whose earnings of $3.50 a day or less provided a bare living. By carrying a few chunks of good ore out in their pocket or lunch bucket they could increase their week's pay to as much as $100. Most of the extra cash went for liquor, girls, and gambling and very little toward clothes or better living conditions. As one miner explained, "When you had muckers working for you and they suddenly started spending freely, it was time to check up on the high-grading."

Grover Kane, whose father came to the Yellow Aster as a miner in the rush of '97 and whose mother ran a boarding house at the junction of Broadway and Rand Street, told of high-grading by the men that boarded at their place. "They never had real high grade at the Yellow Aster, but sometimes they would hit a streak that would run $3 to $4 a pound instead of the usual $1. If a fellow got away with even the dollar stuff, a few pounds would add up to more than his day's wages."

"I was just a kid of ten and had more money to spend than I knew what to do with. Mother would put an extra piece of pie in a boarder's lunch pail, and when he brought it in at night, the handle would hardly hold the pail up, it was that heavy. I'd relieve them of the bucket before they went into the dining room for dinner and dump the contents into a box out back. We got a share, and probably the rest went to pay the fellow's board bill, which was $25 a month. Most of the men were making $75 or less a month. Gold was selling at $20.67 an ounce."

Even $75 a month guaranteed income would have seemed good to some of the men trying to develop their own digs.

Similar to gold boom localities all over the southwest, the Rand Mining District and associated areas had been combed by prospectors hoping that an overlooked canyon, valley, or mountainside would turn out to be as good or better than the Yellow Aster. Old diggings of the 1880s and early 1890s were checked again for indications of gold that had been missed. Men searched and lived on very little or were grubstaked by others who hoped the unexpected would happen and they would benefit.

The old Summit Diggings had been well worked before the Yellow Aster discoverers, Mooers, Singleton, and Burcham arrived in the area. The Diggings were still supporting a few miners when two brothers, Rea and Tony Spangler, started prospecting to the east. The unlikely looking hills they prospected lay near the tufa formation of an ancient lake; earlier prospectors had bypassed the area since there was no indication of gold bearing strata.

In a short journal left by Rea Spangler, he explained that the Spanglers left Tulare County in the fall of 1896. With four mules hitched to a light wagon loaded with feed and 150 gallons of water, they started for the gold area of eastern Kern County, mainly Rand Camp. They crossed the Greenhorn Mountains just before the first snow, dropped down through Kernville and over Walker Pass and rested at Raymond's Coyote Holes Station. They then traveled east along the El Paso Mountains, finally stopping at the Summit Diggings. Prospecting north and east of Rand Camp they worked toward a range of dark hills that stood out against the tan of the desert. It was here they found float showing gold on all sides. The assay ran $45 a ton, and they called their claims Goldpoints.

Returning to their home near Delano, the Spangler boys loaded their 12- by 16-foot frame house on a wagon and with eight mules pulled it over the same route they had traveled some months earlier. The next trip brought the remainder of their mules, hay and grain. As soon as their camp was set up, they started mining. With gads, hand picks, shovels, and jack hammer they developed over 4,000 feet of underground tunnels and drifts. By means of two wagons pulled by 10 mules they hauled their ore to the Garlock mill where they received an average of $70 a ton. Through the years, whether the gold market was good or poor the brothers went on mining until age stopped them.

As investors' money flowed into the Rand area, mine claims sold for fabulous prices; each new owner felt that he could make the mine a bigger producer than the owner before him. An example was the Big Butte, discovered in 1896 by three men listed as Sommers, H. C. Ramie, and Tate Ramie. After very little development, the mine sold for a $17,000 profit for each of the three men, a quick return. Sommers took off his overalls, dressed up, and said that he would never work another day. Three months later he was back working at the Butte for $3 a day; he had never left Randsburg during that three-month period and lost every cent of his $17,000 over the tables.

Rumors kept the gold fever at a high pitch and the locals were not above enlarging on so-called discoveries. The smart city slicker, the wise guy, and the gullible were always targets for fabulous gold stories, which sometimes sounded so credible that even old-time miners were duped.

In Frank Nugent's saloon there was almost a nightly bet on the largest nugget found that day, the loser buying the drinks. Two miners, Avery and Skillings, began to make it a regular thing to challenge anyone in the house on the size of the nugget and its richness. Night after night they got free drinks until Nugent figured there was something underhanded going on and thereafter marked each nugget presented for the bet by using a center punch. After that Avery and Skillings had to pay for as many drinks as they won, and many nights they didn't appear at all.

An amusing incident that many Rand residents wished forgotten was the episode in which Gene Reeves, manager of the Rand Mercantile, thought it would be fun to get a bunch of newcomers excited. One morning about mail time when a number of greenhorns were standing around waiting for their mail, Reeves was seen digging near the post-office steps; suddenly he was examining a good-sized nugget. Word of the discovery spread fast, "The streets of Randsburg are filled with nuggets!" Just imagine, "Everyone had been walking right over a fortune!" Within minutes residents and strangers alike began arriving with picks and shovels. As they started to dig up the street, Reeves tried to pass off his nugget as a big joke, but no one considered it a laughing matter. Gold was gold, and the street was as good a place to mine it as anywhere else.

No one would believe Reeves when he tried to explain that he had just been playing a joke and that the nugget he was supposed to have found near the post-office steps was the one he always carried in his pocket that most of them had previously examined many times. Even the Judge arrived, pick in hand, and was reluctant to accept Reeves' explanation even though he recognized the nugget. There were those who never did appreciate the joke that caught them instead of the ones it was planned for.

For many, the everyday laughs came out of a bottle. When things were going well at the Yellow Aster, the owners kept their men happy by allowing one free drink at any bar on the man's way home from work. At one time there were 33 saloons, most of them with sawdust floors, along Butte Ave. Some of the fellows took advantage of the offer and made up for the nondrinkers by having a free drink in every saloon from Shorty Hillman's at the lower end of town, up to Hattie Brown's. It took them a long time to get home.

Though Randsburg was very much in the public eye, the rest of the area was making history too. Just around the mountain, Johannesburg had been founded. Chauncey Depew and Associates laid out the town site on a half section of land before any residents arrived; thus the community avoided some of the growing pains experienced by Randsburg. Even water, which had been the greatest lack in the Rand Camp, was provided for in Johannesburg's planning stage. Water was piped in from Mountain Springs to the east. An inducement to move to Johannesburg appeared in the Randsburg paper: "Water free to all residents until July 1, 1898, Johannesburg Milling and Water Company." Later Squaw Springs was added to the water system, and Randsburg joined the area served by the water company.

Though the new community considered itself the elite of the mining settlements, it still was subject to the crunching, thudding, and grinding sounds of the Red Dog Mill on the edge of town. Saloons had their brawls and gun play, and drunken miners would go down the street singing just as loud and off key as in any of the other communities in the area.

Beside mining and milling, other projects were under way. Of these the proposed railroad was considered most important. In May 1897 J. M. Beckley of Rochester, New York, Albert Smith of New York City, and A. A. Daugherty of Los Angeles became enthusiastic about developing a railroad to the gold mines of the Rand District. They became the leaders in promoting and investing in the new Randsburg Railroad. The interest and financial possibilities were so good that surveyors were put on the job immediately. Their recommendation was for an inexpensive, straight line north from Kramer Junction (now known as Four Corners) instead of from Mojave, along the valley to Garlock, and up the mountain from there.

Work began October 2, 1897. A train load of laborers came in from Los Angeles as 10 carloads of steel rail arrived at Kramer where the new line would join with the Santa Fe. A reconditioned engine was sent out to haul supplies along the new track as each section was completed.

Construction progressed at top speed the first month; then supplies failed to arrive and all work had to be halted. Twenty-three miles of the 29-mile roadbed were already graded and half of the track laid. The company had promised the Rand area rail service by the first of December, but the supply delay caused them to revise their plans. Passengers could ride from Kramer to St. Elmo and then take a stage the remaining six miles to Johannesburg or seven and a half miles to Randsburg.

The surveyors had marked out a straight line for the track, ignoring working mines and mine claims. In one such case, contractors Ramish and Marsh planned on laying tracks right over a working mine shaft. The miners Webb and Wrem over whose claim the rails were being laid objected and expected the mining laws to protect them. To beat any legal delay the contractors put down the rails during the night. Next morning the miners could not enter the mouth of their shaft. They organized a group of sympathetic miners and planned to tear up the rails the following night, but someone tipped off the contractors. Twelve armed guards stopped the miners and kept them off the property until they finally gave up and went mining elsewhere.

By Christmas the line had reached Johannesburg, and it was expected that the rails would continue on to Randsburg. Albert Smith, investor and general manager of the line, fed the rumor by adding that eventually the road would be extended another 60 miles to serve the Death Valley region. Optimistic promotors assured the public that in time the rails would continue on north to connect with the Carson and Colorado at Keeler. But the Randsburg Railroad reached Johannesburg and never went an inch farther, not even the short distance to Randsburg.

On January 5, 1898, regular trains started operating. One round trip each day gave the impression of ideal transportation into and out of the area. In reality it was a poorly planned schedule and a time-consuming trip missing all connections and requiring long layovers. Twelve days later two daily trains provided through connections with Barstow to the east and Los Angeles to the south. One could leave Johannesburg at 9 p.m. and arrive in Los Angeles at 6:30 the following morning.

To make the Randsburg Railroad a real asset to its investors, president Beckley and his associates formed the Randsburg-Santa Fe Reduction Works, erecting a 50-stamp mill west of Barstow. The plan was to haul and process the Yellow Aster ore. Milling operations started in June 1898 and immediately cut into the amount of ore hauled down to the Garlock mills. And Garlock had already felt the effects of the 30-stamp mill installed right at the Yellow Aster mine site.

Eight months later the Yellow Aster Mining and Milling Company completed an additional mill; this was a big 100-stamp mill. The Barstow mill shut down for lack of ore, and three mills in Garlock were soon forced to do the same.

There were mixed emotions over the Yellow Aster's new mill. Some people feared the loss of the railroad when ore shipping came to a halt; others were elated over the prospect of additional jobs created at the mine. Both interests were suddenly shoved into the background as excitement rippled over a rich gold vein found not too far from the Red Dog Mill. The discovery was to become known as the Phoenix Mine. The discoverer must have been hopeful of its being superior or having lasting immortality from the name, but it never lived up to expectations. Instead, there seemed to be a kind of jinx hanging over it.

Claude Holloway and an 80-year-old prospector known only as Old Bill were two of the first miners working the Phoenix. Both were considered excellent miners, and each considered the other a good, reliable working companion, which was all important in the dangerous business of blasting, tunneling, and drilling. Each morning they spent in mucking out and shoring up, or, in layman's language, they cleaned out what had been blasted loose the day before and then put in timber supports for overhead and walls.

For days the work had gone along well with the gold vein running parallel to the entrance level. The pitch of the vein made considerable difference in the ease of drilling and the amount of support needed before they progressed deeper.

One morning following the usual clean-out, the

two men came out of the hole to rest and eat their lunch in the warmth of the spring sun. It was such a pleasant day that they continued to talk and delayed going back to work. Finally they agreed that they had delayed long enough and kidded about having spring fever.

As the men hurried down the long tunnel in an effort to make up time, Old Bill remarked, "Seems like a lot of dust in the air."

"Don't remember any when we went to lunch," Holloway added, and then suddenly they both came to a halt.

That night in Jack Harrison's saloon they tried to explain just how it felt to suddenly come to the spot where you had been working an hour before and find that a 10-ton boulder had dropped from the retaining wall, burying your tools and footprints. Both believed they would have been crushed to death by the boulder if they had returned to work at their usual time.

After that incident there were miners who refused to work at the Phoenix, believing that the falling boulder was a sign to stay out of the mine. In spite of this, the operation developed into a good-sized mine and mill. Mine accidents happened at most mines in the area and caused little comment, but when the same thing happened at the Phoenix, the old superstition of an ill omen would again gain power and there would be a general exodus of workers. A prediction, often repeated, declared that some day a great catastrophe would no doubt strike and the men working there would be caught in the disaster.

An early resident remarked that it took real courage to work in some of the mines of the area, but it took "guts to survive the winters." He explained that even if one was fortunate enough to have a tent, a small river would run over the floor most of the time, boots were wet through day after day, and clothing and bedding remained cold and soggy. Fuel was scarce and expensive.

The winter of 1897-98 was a bad one, but living conditions were much improved over those of 1895; the miners had tried to survive the cold under makeshift tarps that were far from the comfort of a tent. Wood structures were few but slowly on the increase; they were just shells, and the wind blew through nearly as much as in a tent. The hotels and boarding houses were noted for the cold drafts that air-conditioned each room, and the outdoor sanitary facilities seemed to always be on the windward side.

Coal freighted in by team and wagon helped to alleviate some of the discomfort. In later years few could recall which places of business offered fuel for sale, but people well remembered Tom Duke's wagon drawn by a beautiful white team as he delivered coal to Randsburg, Johannesburg, Summit Diggings, Goler, and Garlock as well as to mine operations scattered along the route. A snowstorm struck about the time a coal strike tied up eastern mines, and the supply in Mojave Desert communities ran low. Lumber hauled in for construction had to be guarded day and night or it would turn into someone's firewood. One stack of lumber, all sawed ready for use in a dance hall framing and underpinning, melted away one night right under the nose of the guard.

Liquor was credited with keeping up the miners' courage. For some, not even liquor helped; they just hitched up their team or caught the next stage out and headed for the warmer valley operations in Goler and Red Rock or quit the area entirely.

Garlock School, 1896. Front row, second from left: J. H. Martin; back row, from left: a Henderson boy, Walter Martin, Ed Mackey, Arthur Martin.
(Courtesy J. H. Martin)

6

As They Saw It

Adults looked at the mining area almost entirely from the standpoint of making a living; their youngsters saw the gold boom communities from a different point of view.

In October 1897, John Cuddeback, freighter and builder, completed a rock house and brought his family to Randsburg. Seven-year-old Grace remembered that her father's first job was building a store in Fiddlers Gulch. Then he started buying wooden buildings in Garlock and moving them to fill the shortage of houses in Randsburg. "Our mother rented them as soon as father could move them up the mountain and set them in place. One four-room house had to be cut into sections before we could move it through the heavy sand of the valley and inch it up the steep grade to Randsburg."

There was no school at the time the Cuddebacks arrived in the mining camp, but by the next fall the Wolfskill store building in Fiddlers Gulch had been acquired by the citizens and moved nearer the center of town, at the head of Butte Avenue. The building was the one Cuddeback had built when he first arrived; its location at the far end of town had proved a poor place for a store or saloon and the owner had been glad to dispose of it without loss.

The first teachers, Miss Mite and Miss Williams, didn't stay long; and neither did those that followed them. A teacher had to be of sturdy stuff to manage children of all ages and sizes, some as large and nearly as old as the teacher herself. The big boys delighted in seeing how difficult they could make life for a young woman.

Grace Cuddeback spoke of one of these incidents as the time Leo Cane dared a Miss Witters to punish him, since he was one of the "big boys." Before he realized what was happening, and in front of the whole school, the teacher flipped him to the ground, held him down with one foot, and proceeded to give him a real whipping.

The mode of dress for schoolgirls was fairly uniform: long woolen dresses in winter protected by a white pinafore; warm weather brought out cotton dresses that were frequently covered by the same white pinafore. Stocking styles called for long dark stockings in winter, short ones in summer. Grace Cuddeback added that, "We always wore a hat, winter or summer, wherever we went."

Grover Kane, who was in Randsburg about the same time as Grace Cuddeback, recalled the school as one large room with a partition in the center separating the two teachers. A Mrs. Higgins had the younger children, which was the age group he was in, and a Mrs. Rader had the older students. The redwood desks were built for six students in a row. "They had the boys on one side and the girls on the other, just so we couldn't tease the girls."

The original old school building, now a home, still stands at the head of Butte Avenue, a picture of early architecture with its gingerbread trim. One of very few frame structures to survive the big fires, the old building served the community until a larger school was built in 1900.

Down in the valley another young person was starting on the road to new experiences. Laura Meeker Cole told how her uncle, Fred Wells, moved her mother, Clara Doty Meeker, and two sisters, Dolly and Pearl, from Hanford to Garlock in 1897. "We traveled by wagon six days, camping at Bakersfield, Caliente, Mojave, and Kane Springs and finally reaching Garlock, where my grandparents, A. J. and Sarah Doty, owned and operated the only two-story hotel in town."

The Doty Hotel was in a class with the two-story hotel in Johannesburg and outclassed it when Garlock, the first community to have such luxury, put in street lights. One of the fine lantern-type street lights stood at the corner of the hotel.

A continuous flow of freight traffic went through the town as the stages brought in miners, investors, and salesmen. The Doty Hotel, offering home-cooked meals and rooms for rent by the night or week, was headquarters for the stage and soon became the hub of Garlock activity.

Seven-year-old Laura went to school down the street and helped at the hotel after school. Though there was a good well just outside the building, water had to be carried in a five-gallon pail for customer use, and it seemed like a never-ending chore before all the hotel needs were cared for and she had time to play.

At that time the Doty Hotel was considered a spacious structure with its ten bedrooms, a parlor, lobby, large dining room, and kitchen; family quarters were across the back of the building. All food came in by freight wagon and most of it was either dried or canned; fresh fruit was a rare treat. The one exception was fresh meat, which they bought three times a week from the Beatty slaughterhouse on the edge of town.

The Doty family, early California pioneers and famed for their oxen and long teams used in sequoia logging days of the 1880s and 1890s, had moved with the tide of activity. By 1896 Charlie and Elmer were freighting supplies from Mojave to the mine camps and ore from the mines to the Garlock mill on the west end of town and the McKernan mill to the east. They were hauling ore some distance, as milling in Garlock was $8 a ton compared to $15 charged by the mill out on Cuddeback Lake; in many instances Cuddeback would have been a far shorter haul. Jack and Elmer Doty, sons of A. J., the hotel owner, were known all over the country for their baseball playing. Elmer was pitcher and Jack was catcher, an almost unbeatable combination.

Laura Meeker, like other children in the desert mining camps, enjoyed riding the burros that hung around town for water and feed. Will Doty, her mother's cousin, who worked in the Golden Queen Mercantile, didn't think it was proper for Laura to wear dresses when she rode a burro. He bought her a pair of overalls, shirt, and shoes like the boys wore and told her she couldn't ride anywhere if she didn't wear them. "It was a real struggle to make myself put on those boy's clothes and at first I would ride on the back streets where I hoped no one would see me. When

I got used to them, I didn't want to wear anything else."

Laura Meeker Cole added, "Children were always making up rhymes and songs in those days, and the one I haven't forgotten went

The men, they grow tall in Garlock
The girls, they grow small
But the men love them all
And they marry in the fall
In Garlock.

Laura's widowed mother married James Curry, and in 1905 they moved to Johannesburg, where he ran a feed and fuel business. Their house was a three-room one moved up from Garlock with three rooms added. Besides being one of the largest, theirs was the only house in town with a bathtub. A small flower garden was carefully tended at the side of the house. She felt that they were fortunate: "Our father had a feed lot and could afford to feed a cow so we had good milk, not the watered down output from the Johannesburg Dairy." Curry also managed the water system for owner C. J. Teagle, taking water in payment. Elsewhere in town water cost 10¢ to water a horse and 50¢ a barrel for household use, but the freighters who used the Curry feed lot could water their stock free.

Johannesburg, 1905. Back row, second from left: Irene (Doty) Phillips, Clara (Doty) Meeker, Mr. Curry, James Curry; middle row: Pearl Meeker, Laura Meeker; front row; Dolly Meeker, Irene Phillips.
(Courtesy Laura Meeker Cole)

When as a young lady Laura became interested in social functions, the Johannesburg Club House was considered a respectable place for the youth of the community. Dances and parties were well supervised there, but all other places were forbidden. "But one Halloween party slipped out from under the supervision. We were playing follow the leader, when suddenly the leader dashed out the door and down the street; everyone followed in the spirit of the evening. I don't recall who the leader was, but he was very daring as he led us into a saloon, wove in and out among the tables and back out onto the street. We got an inside look at all the forbidden spots that night. You should have seen the startled faces as we made a quick tour through a place. Not a thing was said to us by our parents the next day."

Frank Lillard gave a nine-year-old boy's impression of Garlock in 1898 when his parents moved from Los Angeles by wagon to seek their fortune in the gold fields of the Mojave Desert. Z. T. Lillard, the father, expected to make a quick fortune in gold, but his wife Nancy Jane was more realistic and brought along her cooking utensils and all the sugar and flour she could get into the wagon after their few clothes were packed.

"The minute we arrived in Garlock mother set up a camp kitchen near the wagon and started making doughnuts," Frank recalled. "The miners bought them as fast as she made them. The money went to buy more flour and lard; soon the doughnuts paid for a stove, and she added fresh bread, then pies." The baking paid for a one-room shack, then another room for a kitchen; more rooms were added as there was money to pay for the lumber, and the business grew into a modest hotel where Mrs. Lillard and her two daughters "Tom" and "Dick" served home-cooked meals. Mr. Lillard ran the front desk and discussed mining investments.

The cooking was hard work and long hours with a fire going full blast regardless of the temperature outside. In later years Mrs. Lillard confided to her son that during the time they ran the hotel in Garlock, she never got over four hours' sleep a night as the pies and bread had to be baked before starting the early breakfast for the miners and mill hands who were regular boarders. She cooked about 300 meals a day to serve approximately 100 men, not counting the transients who stayed a day or more. The 100 lunches were packed before breakfast, ready for the men to take as they left the dining room after their morning meal. "Each man had his own tin lunch pail; sandwiches and pie went in the bottom and coffee in the double lid."

Of his own experiences, Frank Lillard remembered most meeting the stages, especially the one that brought in the *San Francisco Bulletin*. He felt quite the business man when the driver would yell, "Here's your papers, boy!" Deliveries were fun, for he knew everyone in town, but it was difficult to sell extras to the strangers who stayed at the hotel, "Anyway Dad usually took care of that as he visited with everyone that stopped; spent most of the time in the sitting room or on the porch smoking a cigar and talking mining."

Hot spells sometimes hit suddenly or unusually early and before anyone realized what had happened the mound-shaped chocolates in a tall glass jar on the hotel desk would lose their chocolate coating and only the cream centers would keep their shape. "It was a drippy, sticky mess, but wonderful," Lillard delighted in the memory. "I got to eat the whole batch, right down to the liquid chocolate on the bottom." Since he and the other boys often made up chants about the ball team, the freighters, and everything else that interested them, it is likely that the chocolates were so honored, but in later years the only rhyme he could remember was

> Potatoes grow small in Garlock
> We eat them skin and all.

When the Yellow Aster Mine at Randsburg built a pumping station at a well just east of Garlock at the entrance of Goler Canyon and boosted water up to the new 100-stamp mill, workmen began to leave Garlock. The small mines of the area were not supplying enough ore to keep all the mills running, and business in general started dropping off. The Lillards were considering shutting down their hotel when Charles Burcham, one of the Yellow Aster discoverers, came by and remarked, "Zeke, I'm sorry we took your town away, but I need a manager for the company hotel, so how about moving to Randsburg?"

The move was made, and Mrs. Lillard said that that was the end of cooking; she had done her share and it was time Zeke carried his part. But when Burcham caught the hotel cook handing a ham out the window to some strikers, he talked her into temporarily filling the fired cook's job. Her son recalled, "She went back to cooking; most people did what Mr. Burcham asked; he had a nice way about him."

Harrison Doyle, another boy of the early Rand Camp, lived with his family in a dugout purchased by his dad for $40 in gold pieces. The family of six lived there several months while the father prospected. Young Doyle increased the income by selling empty whiskey bottles to the saloons and ran errands for the girls in the canvas and tar paper dwellings known as the floozie barns. In writing of his boyhood in Randsburg from 1896 to 1900 he described the scene from their dugout home

> . . . about a block behind the main street. Even at this early date there were three saloons, an assay office, the St. Elmo Hotel, and the Elite Dance Hall. Between our dugout and the back of the Elite was the canvas and tarpaper red-light row. And between us and the St. Elmo stood the latter's board and batten outhouse, one end for the women, the other for the men.
>
> The Elite was a board-on-end affair, tinder dry. It had a tinny piano, dance floor and small stage.
>
> I used to wander barefoot into the St. Elmo to look at the fist-sized nugget and other specimens on the counter, and to run errands for the hotel clerk.

It was not until later that he found out that the fist-sized nugget was a gilded chunk of lead to fool the tenderfoot.

In addition to the camp impressions made upon the young members of the mining communities, there was the traveling photographer who not only left a valuable treasure of pictures but was a part of those early memories.

Young Bill Wilkerson was helping his father at their Mattie Mine the day the Los Angeles photographer toured the area, photographing families, mining operations, and individuals. "When the photographer was around, everyone wanted pictures made so that distant relatives or folks back home could see how the children had grown and just what the mining setup looked like."

"Father came up out of the mine, and all activity came to a halt as the photographer and his assistants sighted here and there, searching for the most inclusive view. We were all lined up for the picture, but by the time he decided just how he would take the photo, the light had changed. There was another long wait as the camera and all the other paraphernalia were moved and rearranged. I was the youngest in the family and had never had my picture taken. It was exciting, but I grew tired of waiting for the fellow to quit changing his mind.

"Finally the camera was in the right place; we were in the right place; he draped a cloth over his head, raised it, peered out and then covered his head again. 'Ready!' When he peeled the cloth off his head and announced, 'That's it,' I was so disappointed. The fellow just picked up his stuff and left and there wasn't any picture, though father did pay him some money. I just couldn't understand why he hadn't left a picture or why my father had given him money after all the time he had wasted while he kept us out of the mine."

Days later when the incident was forgotten, a photo arrived by mail. Wilkerson added, "There we all were, just like the day the photographer fellow took so much time and didn't seem to do anything."

The traveling photographer did a surprisingly good job considering the little that was known then about the process and the unwieldy equipment used. Thanks to the ones who traveled in a covered wagon with their studio and living quarters under one roof, the Rand Mining area's earliest activity was recorded on glass plates and then painstakingly transferred to paper that would be glued to a cardboard back. Light, moisture, stains, and time have left their mark, yet the old photos continue to be the visual aid to understanding a vanished period.

Traveling photographer had living quarters and studio in his covered wagon. Thanks to such men who worked under hardships and with primitive equipment, we have a record of the people, the mines, and the styles of the Rand Mining District. *(Author's collection)*

Methodist Church fire.

7

Fire

Fire! The very word struck fear in the minds of most miners who had seen camps all over the southwest go up in smoke in a matter of minutes. It had worried Rand Camp residents as more tents crowded against each other, wood business buildings almost touched, and no adequate source of water had been developed.

The *Randsburg Miner*, May 7, 1898, gave the date of the first fire as December 23, 1897, but told nothing of the extent of that damage. Little attention was given that fire, so it is possible that it was quickly gotten under control.

Less than a month later, on January 18, 1898, the big fire roared through the camp. On the following day the *Los Angeles Times* ran the story:

> Rand street is in ashes. Every building is burned between Broadway and Staley Ave. on the east side, and on the west side taking down to the Elite Theatre, down below Nelson's realestate office. On Broadway four buildings were burned on the north side, stopping at Wagner's saloon. The buildings on the south side of Broadway and on the north side of Butte Ave. opposite were scorched and were many times on fire, but were saved through the heroic efforts of the firemen and others present.
>
> The alarm was sounded a little after midnight, the fire being first discovered in the rear end of the Mojave Saloon on Rand Street. It soon got beyond control and the adjacent buildings were wrapped in flames. The night was perfectly still, but the dry wooden buildings burned like tinder and the flames and smoke, mounted hundreds of feet skyward, made a grand but awful sight. Men, women and children filled the streets, everybody near the fire being intent on saving their houses and furniture. Part of the contents and furniture were removed from most of the buildings, except in the St. Elmo Hotel, where everything was lost, the guests escaping with their clothing only.

The post office was burned with all fixtures, but the mail was saved, being hastily thrown into bags and carried to a place of safety.

For two long hours the suspense was terrible, there being no certainty that any part of the town would be saved. Many people remained up all night watching the fires or protecting their goods scattered everywhere about the burned district. Today many are making preparations to rebuild, and the people, notwithstanding their losses are cheerful and hopeful. Other buildings are being moved onto the vacant lots and some firms will resume business tomorrow.

The primary firms burned out are: Hammond's Grocery, Postoffice, Wilson's General Merchandise store, Southern Pacific office, St. Elmo Hotel, Elite Theatre, White Fawn Saloon, Crawford's Saloon, Hooper's Merchandising Store, Postal Telegraph office, Hemstead's Drug Store and numerous realestate and assayer's offices, small stores, saloons and bakery.

Supplies were rushed into the destitute camp, and merchants set up makeshift tents and lean-tos in order to supply the immediate needs of the residents. Some decided to take a little more time in rebuilding and have the new structure made of adobe as protection against another fire.

Volunteer firemen were organized and seriously trained. The first "fire engine" was a dilapidated wagon to which two barrels of water were fastened on either side. A fireman stood on the platform at the back, and one horse pulled the equipment along the street. Eventually two fire companies were organized and had regular weekly drills to perfect methods and speed to reach any part of town. Fire plugs and pipelines were placed at advantageous locations in readiness for water when the town would cease to be a dry camp. Johannesburg on the other side of the mountain had water in limited amounts, but had agreed to furnish water in case there was another fire. The big flaw in the plan was that the pipeline was not long enough to join the two towns.

One small youngster later remarked that she could remember very little about the fire as, "My mother, like a lot of others, was afraid the stored dynamite, which was all over town, would explode; so the children were quickly taken into the surrounding hills, and we missed all the excitement."

Five months later, almost to the day, the town was leveled again. The *Los Angeles Times* reported:

By direct wire.
Randsburg, May 6, 1898

Another terrible fire in Randsburg wiped out all the remaining part of the town and not a building is left from the corner of Staley Ave. up Butte Ave. to the Russ House on the other side, and but four buildings are left on the south side of Broadway and three on the north.

The fire started in George Glutter's residence on Butte Ave. about 3 P.M. and was well underway before the alarm was given. A strong wind was blowing from the west up the street.

In an incredibly short time it spread both ways, crossing the street and taking the Express and Telegraph office, together with the bank, which only moved in yesterday.

It crossed over Broadway, taking all on that side. It then raged fiercely up the avenue on both sides, making heat so intense that no one could go near it. On the south side of the street Ord's hardware was the last building burned, while on the other it stopped at the Russ House.

Several persons were hurt. Miss Griffin was severely burned and her life is doubtful. Her clothing caught fire and she was terribly burned about the face.

The business loss is enormous. But one general store is left in the place, and there is a general feeling of despondency. The principal losses are the Orpheus Theatre, Anderson's store, the San Joaquin Supply Company, Jo Endl's Saloon, the Occidental Hotel, the Phillips House, the Randsburg Lumber Co. which is still burning tonight, Jones' dry goods store, Miller's Drug Store, the Bouquet Saloon, McPhersons' Saloon, the French Restaurant and saloon, the livery stable, two laundries, Varney's Butcher Shop, Desert Queen Saloon, Lee's Shoe Store, Pool's Blacksmith Shop, the Orpheus Restaurant, the Los Angeles Hotel, Mansion House, Hike's secondhand store, Dineen's General Store. Anderson had a heavy stock and had just moved into his supposedly fireproof building.

The loss cannot be estimated but it is very great. Thirty or forty dwelling places have burned and tonight every unburned house in town is crowded. Many men had given up their bed tonight to accommodate women and children.

The fire was finally stopped by the free use of dynamite, blowing up all small buildings. Men are on guard to save the remains of the town. The Orpheus Theatre reopened tonight in the Skating Rink, and many if not all the business houses will be rebuilt.

The *Miner*'s (newspaper) office was three times on fire, but heroic work saved the building and contents with but little damage. The fire reached Wagner's Saloon and stopped. The peculiarity of this is the fact that the last fire came as far as this building at the other side and stopped, leaving him as it has this time.

The following day the *Randsburg Miner* gave more insight into the tragedy. The fire that started at George Glutter's was caused by soup boiling over on a two-burner gas stove. The soup put out the flame of one burner, but the gas kept flowing and reached the other burner to cause an explosion. It was believed the whole town would have gone up in flames if it had not been for the work of the "powder brigade" who blew up building after building in the path of the fire. Josephine Griffin, the one person to be seriously injured, was badly burned when the tent she lived in at the rear of her restaurant caught fire and fell on her.

Randsburg, 1898. Celebrating new water system.
(Courtesy Kern County Museum at Randsburg)

Fire loss was estimated at $95,400, with A. Anderson, who had just rebuilt after the January fire, losing the largest amount, approximately $20,000. Jos Petrich who had rebuilt following the December and January fires was wiped out again with a $10,000 loss; other three-time losers were W. H. Hiveren, E. M. Dineen, J. G. Endl, and Fred Hinche. The list carried the names of 62 businesses and individuals.

The *Miner* ran a column of "What some of the boys lost in the fire" in an effort to bring a smile to the sad community:

Jo P. Carroll lost confidence and went to bed after the fire.

J.W. McLeod lost action and has been stiff as a poker ever since the eventful night.

Peter Klienfelter lost an opportunity. It was one of the extraordinary kind, was gilt edged and bound in leather.

Harry McKean lost his appetite. It was a good one and Harry will pay a liberal reward for its return in uninjured condition.

J.T. Keck lost his whiskers. They were not burned in the fire, but a man who attempted to pull his leg, made a mistake and got hold of the whiskers.

Clyde Kuffel lost part of his scalp. He did not recognize the man who stepped on his head but knew him by his track which he found on the hat the next morning. The hat was not seriously damaged; it was old hat.

A. J. Petter started rebuilding his assay office the morning after the fire. George Toedt moved his home down to his business location as soon as the ashes could be cleared away, and Mrs. Toedt reopened her millinery business the following day. S. J. Montgomery, who had merchandise stored in a fireproof basement, reopened his business with the undamaged stock. When J. M. Crawford's big safe was opened, he was delighted to find the contents uninjured except for the oxidation of the silver coins.

W. C. (Bill) Wilkerson, mill man at the Yellow Aster, later told his impressions of the fire. "A man came through the mine calling out, 'The town is burning up! Everybody ordered out to fight fire.' When we reached the open, I remember seeing the theater just catching fire; the flames quickly burned half the name off. By the time we got down the hill, the theater was completely gone.

"There was very little water, so some buildings were dynamited to save others. If the building was small, a number of men would get hold of it and roll and drag it out of the line of fire. The Yellow Aster paid every man his time though he was out fighting the fire. Everyone in the mine and mill fought to save the town except a Swiss named Henry Bapp, who was stone deaf from a mine explosion; he didn't hear the alarm and went on working down in the mine all by himself."

By November of 1898 the Citizens Committee, in cooperation with the Volunteer Fire Department had passed some fire ordinances. Section 16 read:

> In no house, building or tent, situated on Butte Ave. or Broadway in the town of Randsburg, from the intersection of said streets with Rand St. to the westerly line of the Jim Crow Quartz Mine Claim and for a distance of 100 ft. on either side of said streets shall there be located or used any gasoline stove, unless with written permission of the Fire Commission of the Randsburg Fire District.

In addition to the new regulations to prevent fires, John C. Quinn proposed constructing a tank on Kenyon Hill that would hold 100,000 gallons of water, enough to keep a 2-inch nozzle in operation for four hours. It would mean raising approximately $3,000 from the citizens to start the project. After that there would be the continuing expense of hiring teams to haul water to replace the 1,000 gallons a month that would be lost by evaporation. Captain Colson of Squaw Springs had offered 400 gallons from his wells, and the remainder could be brought in tank cars by railroad. The Citizens Committee took the suggestion under advisement.

The lack of water for milling, for fighting fires, or even for household use was a continuing problem. That same fall the Yellow Aster purchased the Skillings water wells six miles from Johannesburg, a source that could be relied upon to run a 30-stamp mill that was under construction. The mill would handle approximately 140 tons of ore per day, producing $50,000 in gold per month.

Although the Skillings wells were supposed to be adequate, the Yellow Aster made its position more secure by taking a lease on the historic old water stop, Desert Springs, near the present Cantil. If the two water sources were good, a third would be even better, and the Benson well two and a half miles east of Garlock was added. Both because the title was questionable and because this hand-dug well, 400 feet deep, had been poorly timbered, the Yellow Aster sank a new shaft a short distance from the original well and got a larger flow of water, averaging 125,000 gallons every 24 hours. But as yet there was no means of pushing that water up to the mill nor to the town, and the lack of water continued to hamper fire prevention and fire fighting efforts.

Down in the valley, Garlock too was having fires, but with an adequate water supply most of them were quickly contained. A boy, J. H. Martin, witnessed one fire in which one building burned to the ground but no other structures were damaged. A one-armed stranger had arrived on the early stage and was going around town, offering to wrestle any two-armed men in town. The betting started, and by night all the men went down to the saloon to watch the free show and place a few dollars on their local man.

Martin recalled that, "Kids weren't allowed in the saloon; so we stood outside and peered in the windows. It looked like the match was about even and the fight was getting exciting, when one of them flung the other high enough for his arm to accidentally hit the hanging lamp. There was a crash, then sudden darkness, followed by an explosion that threw us off our feet but away from the building. By then the place was in flames. It was everyone for himself, and men were running in all directions."

It wasn't until after the saloon was in ashes and the other buildings were safe, that anyone thought of the one-armed man. It was Martin's mother who found the man wandering around, burned from head to foot. Taking him into their one-room adobe home she administered home treatment while someone was sent for Doc Wright. Doc did what he could and then told them that the man couldn't possibly live for long. Weeks later the fellow walked off without even a thank you for the day and night nursing care that Mrs. Martin had given him.

That same year another explosion was etched on the memory of another boy living in Garlock. Early on a cold winter morning an explosion rocked the town, and though dynamite blasts were not unusual, this one seemed to be just outside everyone's door. People poured out of the buildings, each asking the other what had happened.

Brigham, a recently arrived miner, had been having difficulty getting his explosives to function properly on cold mornings, so he had put the frost-covered dynamite in the kitchen oven to warm. Unaware of the contents of her oven, Mrs. Brigham had stoked the fire in preparation for baking a cake. As she opened the door to place the cake in the oven, the whole stove exploded.

As her son, Leslie Brigham, recalled the tragedy, "She lived five hours with a stove lid embedded in her chest and her face gone."

Old-timers agreed that the incident was the worst in the history of Garlock.

John Singleton, far right, joins friends in a ride through Singleton's Court and adjoining wealthy estates along Figueroa Street in Los Angeles. *(Author's collection)*

8

Golden Platter

"Nowhere in the world can such phenomenal success be shown in so short a time as in the great bonanza of the Mojave Desert. Harassed by litigation, persecuted by legal blackmailers, the owners of the Yellow Aster mines have moved forward, overcoming all obstacles, until the grand success of today which gives promise of the opening of another Comstock, making Randsburg another Virginia City," thus said the local newspaper, *Randsburg Miner*, November 17, 1900.

The thing the newspaper didn't tell was that except for the operating expenses of the great mine, very little of the wealth was spent in the area. Within three years the three near-starving miners had become wealthy, and less and less was seen of them as their income increased.

F. M. Mooers, the ex-newspaper man, was the first to leave. In 1898 because of poor health he moved from the mining community to a luxurious residence on Bonnie Brea Street in Los Angeles, harassed and in the process of divorcing his wife, who would not live with him even after an agreement that gave her a million dollars' worth of stock in the Yellow Aster. Much of the time Mooers was too ill to take care of business matters, though his job as vice president of the Yellow Aster Mining and Milling Company was not too demanding. He brought his aged mother and a brother from the East; the brother was to look after the financial affairs.

John Singleton, the former carpenter and millwright, president and general manager of the Yellow Aster, soon followed Mooers to Los Angeles and built Singleton Court, a showplace of its time. The magnificent mansion, set in a three-acre estate at Adams and Figueroa, was equaled or surpassed by the elaborate stable.

Singleton's stable was considered the most costly in Los Angeles. It was built of bricks held together with cement in which gold-bearing rock from the Yellow Aster was embedded. The stalls, made of mahogany, with hardwood mangers, housed thoroughbred racehorses. The second story of the building was topped by a 100-foot bell tower that also supported an ornate clock with four dial faces, each 5 feet in diameter.

The massive mansion, set in a subtropical garden of exotic trees and shrubs imported from all over the world, preserved its seclusion behind a wrought-iron fence. On gateposts were embedded letters reading "Singleton Court." This was a place of fashionable and elaborate parties until it burned in the early 1900s. The stable remained a landmark, converted to the use of the Orthopedic Hospital until it was demolished in 1967.

Charles Burcham, the cattleman who became vice president of the Yellow Aster, and Rose, the doctor who was secretary of the company, did keep a small home in Randsburg for some years. In 1902 they purchased an elaborate home at the corner of Burlington Avenue and 7th Street in Los Angeles. The house, built in the Italian style, had a large hall and numerous drawing rooms where Dr. Burcham held large receptions when she was in town. She was highly regarded in the social world and famed for entertaining outstanding foreign visitors, business men, and educators. Travel to Europe was expected of the wealthy, and Dr. Rose made lengthy visits to Italy and Paris.

Charles Burcham was in and out of Randsburg and Los Angeles but spent most of his time in San Francisco when not pursuing mining interests throughout California and Nevada. His investments continued in mining; Dr. Rose put much of her returns from the one half of one third interest into property and orange groves of the San Bernardino area.

None of the discoverers of the fabulously rich Yellow Aster properties left anything of substance to the community that had produced the wealth. The only remaining monument is the dust from the great cyanide dump that sweeps over town in a cloud of silt on windy days. But although the wealth was spent elsewhere, the Rand area went on making its own history as individuals played their particular roles.

The Randsburg Citizens Committee continued to work for a better place to live; the Fire Commission passed more decrees for fire prevention requiring stovepipe insulation and ladders to all houses, and even regulating the use of inflammable oils in kindling fires. The town ordered 300 feet of first-class fire hose, and the *Miner* gave front-page recognition to those who observed the Commission's recommendations. The Russ House had erected a ladder "clear to the top of the roof," as had the theater. "Mr. Pearson not only erected a ladder to his place, but also placed a bowl of water on the roof to be used in case of emergency." Even Deputy Assessor J. Kelly was getting cooperation, with practically everyone paying their taxes.

Local phones were installed, connecting the larger mines and the towns of Johannesburg and Randsburg. Telegraph service made it possible to communicate with most of the business world. Wells Fargo took care of the money and gold flowing in and out of the district, while the Methodist and Episcopal churches and a commuting priest took care of the souls.

Transportation was no longer a problem; daily stages left Randsburg at 7:30 a.m., arriving in Mojave at 3 p.m., connecting with the Southern Pacific Railroad to either San Francisco or Los Angeles. A stage left the Johannesburg Hotel at 7 a.m., Monday, Wednesday, and Friday for Ballarat and arrived back at 8 p.m. Tuesday, Thursday, and Saturday. The Randsburg Railroad had a daily train from Johannesburg that made connections with the Santa Fe at Kramer for points east or for Los Angeles.

On May 26, 1900, word reached the Rand District that F. M. Mooers had passed away of a heart attack at the age of 52, in a New York hospital. He left his widow Frances Demorest Mooers, whom he had married in 1872, and a 27-year-old son, Edward D. Mooers. Mooers had been in Randsburg only a month earlier and had hired a rig to go the six miles to Summit Diggings where he, Singleton, and Burcham had made their decision to explore Rand Mountain in 1895.

Immediately after Mooers' death, a legal battle started over the fortune he left. The wife, from whom he had been separated for years, contested the will made a month before his death, which named his brother Charles as executor. The property was left to his one son, three brothers, his mother, and a niece, with one sixth to his wife. Mrs. Mooers, who had started divorce proceedings in 1898, now felt that one sixth of the property was not sufficient for "a surviving widow." She went to court to contest the will, claiming that Mooers' family had coerced him into leaving the majority of his fortune to them and that Mooers' mind was weakened by strong drink and illness.

In turn the brother, Charles, proved that he had left a good job in New York to comply with Mooers' request to come to California to handle the business affairs, and that Mooers had already given the wife Yellow Aster stock worth a million dollars. The published figures concerning the estate placed his worth at approximately $780,000, with 15,301 Yellow Aster shares valued at $765,000, the home on La Brea in Los Angeles at $10,000, personal effects $2,500, and annual revenue from the mining stock running near $30,000.

Mining and everything else took a back seat that July in 1900 after Johannesburg offered a challenge in the local paper:

> The patriotic citizens of Johannesburg are not to be outdone in celebrating the fourth. They do not believe in half measures on occasions of this kind and accordingly they have prepared a program which is enough to turn Garlock or any other old desert town, green with envy. Among the features of the day will be the Great and Only Sam Beers, the Renowned Fire Wizard, imported at enormous expense; there will be a Great Sham Battle between the crews of the Red Dog and Pinmore on their respective tailing ponds. S. H. Fairchild will exhibit his trick mules, Teagle Bros. will distribute a car of groceries. Altogether a mammoth celebration is promised.

Randsburg listed their activities, too: a 12-blast dynamite salute at sunrise and a program similar to the first celebration in 1867, with the addition of a handsome-baby contest, an ice cream race, and horse races.

At 10 a.m. an "imposing parade" proceeded from Miner's Union Hall, down Rand Street, and up Butte, with a counter march down Broadway to the Union Hall. Dr. Nichols was Chief Marshal, and the Rand Cornet Band in a decorated vehicle headed the parade, playing all the way.

On the next float, drawn by four horses driven by Al Houser, rode 60 children, mostly girls dressed in white. With its celebration not starting till later, Johannesburg was able to participate in the Randsburg parade: "The Johannesburg coach, lavishly decorated and filled with beautiful women from our neighbor town, was drawn by six magnificent horses. On top was a black footman with a horn and seated beside him was a little girl dressed in white with long gauze wings representing a fairy. The coach was driven by Miller of the Johannesburg Hotel."

Eagles' Fourth of July Banquet.
(Courtesy Emma Beherns)

A decorated surrey driven by Mrs. J. W. Kelly was filled with ladies of the Glee Club; then came the Ladies Degree of Honor with banners. The Price and Hopper business was represented by a white fawn in a decorated canopy-topped wagon. The Fire Commissioner marched in front of the two brightly festooned hose carts drawn by members of the volunteer companies. A long column of Miners Union members, carrying silk flags and banners, completed the parade.

Back at the hall the morning exercises of orations, readings, and singing were followed by contests, a ball game, a dinner served by the Eagles, fireworks, and an elegant ball. The expenses involved in such a celebration were reported as parade, $26.50; printing, $18.00; literary exercises, $5.00; fireworks, $49.20; contest prizes, $73.50; decorations, $24.75; horse race, $40.00; total, $236.75.

After the excitement of the Fourth of July celebration, the summer heat moved in, and life settled into a routine that was unbroken until promotional material started pulling new people into the area again. The *Randsburg Miner* in a special issue dated November 17, 1900, recommended the climate, claiming that children went barefoot year 'round in the mild temperatures (no mention of snowstorms) and that the moral climate was far above that of most mining areas.

According to the *Miner*, there would soon be a need for additional miners with the coming of the Yellow Aster's proposed 100-stamp mill. Labor would be represented through Miners Union No. 44, which owned its own hall and used it for community activities as well as weekly meetings.

Extensive space was given to summarizing the Rand mining picture to that date, stating that so far $3,000,000 in gold had been taken out. The status of the leading mines was given, reviewing the history of the Yellow Aster and stating that the mine was a good operation paying dividends to $10,000 a month when improvements were being made and double that amount when expenses were less. The income and number of employees could not be stepped up under present conditions, but the company had obtained added water and a new 100-stamp mill was being built just 400 yards from the old 30-stamp mill.

The new mill would cost approximately $100,000, and ore now in sight would keep 130 stamps running constantly for the next 12 to 15 years.

The new pumping system built to boost water up to the mill was also an extensive development. The boiler house, an ell attached to the main engine building, had a floor excavated to a 4-foot depth and filled with 300 tons of cement. The Dean Power pump, driven by cut gear direct from the crankshaft to engine had a capacity of 200,000 gallons every 24 hours, against an 800-pound pressure. The pump, which weighed 15 tons, came in one piece and had to be hauled from the Johannesburg railhead to the well east of Garlock. A smaller pump would raise the water 400 feet out of the well and into reservoirs; from there the big pump would drive the water through a 6-inch pipe, up an elevation of 1,380 feet to the mill seven miles away and a 100,000 gallon reservoir. Listed were those in administrative positions associated with the Yellow Aster: John Singleton, President and General Manager, Charles A. Burcham, Vice President, Rose Burcham, Secretary, C. E. Van Meter, Chief Engineer, Charles Crusoe, Bookkeeper, and S. P. Mahan, Superintendent.

A review of the other large operations also appeared in the *Miner*'s special promotional issue:

The Baltic: One of the first properties worked in the Stringer District. Main shaft 120 ft. considerable drifting. In past three years milled $40,000 in gold. Averaged $100 per ton down, dump average $5.00 per ton. $12 per ton ore at present. Owned by Messers Stephen V. Childs, Frank Griffin, Frank Hicks, 1/3 by C. H. Wynn & Sons. 5 stamp mill.

Gold Coin: M.T. Mann discovered 1895, $15,000 first year, 200 ft. shaft. Drifts at 100 and 150 ft. level. First year averaged $64. per ton, 18 inch vein. Two miles from Randsburg on ground worked by dry washers. $40,000 to date, waste dump screened short time ago produced $5.00 to $11.00 per ton.

The Winnie: Ed Hammond discoverer 1895 in Stringer Dist. Sold to Charles Koehn 1896, worked by leasers, milled at Kane Springs. Mostly $60 to $100 per ton, at 200 ft drift level, $200. per ton ore. Free milling, easy to handle. Leasers $4,000 lately.

Merton Mining and Milling: four miles from Randsburg, discovered Aug. 1896. J. J. Donovan, C. L. Adams. Locally known as the **Buckboard**. Main shaft 350 ft. Drifts at 100, 150, 200, 250 and 350 ft. levels. 10 ft. vein, 1100 ft. of development, $15,000 to date.

Sunshine: T.W. Atkinson, 1896, Main shaft 110 ft., drift 100 ft. Being 2 inches to 18 inches. $100 per ton, $20,000 so far.

New Dry Diggings: or Black Nugget Mine, $7,000 in last 2 months with 2 drywasher. Now 50 men working for $15.00 per day.

Butte Mine: located 1896, bought in 1899 by the Butte Lode Mining Co. Work now a mile under ground. Working 12 shafts, shallow one 50 ft., deepest 325 ft. $250,000 in gold to date. Last two millings brought $9,300. Cost $4,800 for milling and wages. Value $58 per ton, later cyaniding of dump will add $7 to $8 per ton.

Kenyon: located Feb. 1896, almost in town of Randsburg. Not worked steadily. Main shaft 350 ft. and drifting 70 ft, another 150 ft drift shows rich ore but very little is publicly known about it. In 1000 ft development work, output estimated at $160,000. Equipped with hoist, 2 stamp mill with cyanide plant connected. Mill has 8 ton capacity every 24 hours. Famous Wedge Mine adjoins, is supposed that the Wedge vein runs into the Kenyon further down.

Sidney Group: 10 claims southeast of Gray Eagle Group owned by White Brothers, 5 mi. from Randsburg, 7 mi. by wagon road. The Sidney located 1897, 120 ft shaft, adjoining Sidney Annex of 1898. Owner A.C. White. 2 shafts 45 & 30 ft., 35 ft drift, milled 100 ton. Average $35 per ton, now milling at $15 to $65 per ton, plan more development.

Annex: S.J. Montgomery and Barney Maginnis, $10,000 so far.

The same 1900 issue of the *Miner* presented a picture of the businesses in operation at that time, leaving no doubt in the reader's mind that all his wants could be supplied by the variety of firms.

The Great San Joaquin Valley Supply Co.
established 1896, Wm. Hiveren
burned in '98, rebuilt
Best Liquor and Cigars
J.P. Carroll, news correspondent, in charge in day time, keeps bulletin board with latest news.

C. Asher Co. General Merchandise
large double fireproof concrete building, with basement.

White Fawn Saloon, J.R. Price, established 1896, on Rand St, rebuilt after fires

G.W. Lloyd oil, fuel, hauling. Haul for Yellow Aster

Miller's Drug Store, N. N. Miller, burned in '98

Wilson House C. A. Wilson, good meals

Brewing and Bottling Works, Lloyd and Heydlouff handle Maier and Zobelein's Beer
ice year round

French Restaurant since 1896, John Tomicich

Books and News Tom McCarthy, oldest business (not in age man) man in town, all daily papers, candies, nuts, tobacco, cigars.

Jewelry J.H. Pearson

Desert Queen Joe Monoghan, been here since the camp started.

R.W. Osborne For the Best Shoe Mending on Earth Patronize the old man Osborne at his old stand, Butte Ave.

Home Supply stone building, large cold storage refrigerator, fresh vegetables and meat. Lloyd and Seebold

Louie's Place Wife is fond of pets, Belgian Hares, hi-grade chickens, great dane "Tiger", parrot amuses people on the street

Up to Date Store T.J. Kennedy, were in business in Garlock, now business fronting Rand and Butte St.

Realestate and Mines T.W. Weeks
also agent for non-resident property owners.

Many businesses that advertised elsewhere in the same issue were not mentioned in the preceding list, which probably cost as much as an ad and gave less information. Box advertisements, old bills, and receipts were more revealing:

Hotel Johannesburg
2 blocks from Depot
Large two story building, 40 elegant rooms
All modern conveniences, electric bell service, a ladies parlor, reading room, office and bar. Big airy dining room, seats 60 people. European and American plan.
Telegraph service

Randsburg Oil and Fuel Co, 1898
Oil, Gasoline, Distillates, Coal
Office next door to the Skating Rink

F.A. Morgan Company
Dealer in Coal Oil, Gasoline, Lubricating oil and compounds, also carry genuine Black Diamond Coal
Gasoline and coal furnished to stamp mills and hoisting works at lowest prices. We own and operate our own tank cars.
Delivery wagon on the street all the time.
Office and warehouse upper Butte Ave.
Also run passenger and baggage wagons to and from all trains.

Johannesburg Reduction Works
Frank Griffith, F. S. Hicks, J. W. Montgomery
Ores reduced by stamp mills, concentrators and cyanide plant.
Ores and bullion purchased.
Mines developed, bought, and sold.

Tehachapi Milling and Trading Co.
Lumber yards:
Tehachapi
Mojave
West Palmdale
Garlock
Randsburg
Lancaster
Johannesburg
Ballarat

Randsburg Drug Store
Drugs, Patent Medicines, Toilet Articles
N. N. Miller Prop.
Fancy Goods, Cigars, Tobacco

"Dad" Staley's ad read simply "The Village Blacksmith," telling nothing of the giant of a man, weighing 300 pounds and tough as nails. Dad never wore shoes on the job; if a red hot splinter of metal or cinder fell from the forge, he would kick it aside with his bare foot. He attributed his robust health to a beer diet. When thirsty from the heat and work he would send out for a 10¢ bucket of beer; his philosophy: "You can be tough as nails and still be a gentleman."

As 1900 came to a close, the residents of Randsburg were complimenting themselves on getting through the year without a major fire, but they were too early with their assumption. At 8 a.m. on the 28th of December an explosion in the kitchen of the Southern Hotel set fire to the building and in minutes it and the building next to it lay in ashes. The explosion had come from a gasoline stove; the landlady and cook, Mrs. Wilson, had been filling the tank. The stove was in direct violation of the 1898 fire ordinance that forbade gasoline stoves in the business area. Immediately following the fire, the Fire Commission served notice that the order would be enforced and all such stoves had to be replaced by January 1, 1901.

The Southern Hotel fire was one of the first large fires to be contained before it wiped out a major part of the town. By this time pipelines reached to Johannesburg; so there was water for a bucket brigade to wet down adjoining buildings. When word of the Randsburg fire came by telephone, Johannesburg shut off all local water connections so that there would be an increase in water pressure, but there was still not enough pressure for the fire hoses. Although the hotel was a total loss and guests lost clothing, no one was injured. Everyone was able to escape from the breakfast table before the fire reached the dining area.

Ironically, that same week the local paper had carried a statement, which became familiar from much repetition through the years, "Water is again getting a little scarce and the Water Company has begun development work again at Squaw Springs."

Another item, in a lighter vein, whether a spoof at Johannesburg or carrying some hidden meaning for the readers of that period, read:

> They have a sensation at Johannesburg in the shape of a sure enough ghost. His principal place of business seems to be at the depot. He is an unhappy fellow as his presence is made known by long drawn out groans as though he was in agony. The Editor of the Miner heard a few of his efforts before the train started last evening, but had no desire to investigate further. If there are any developments we will report them next week.

Gold continued to be the spark that kept the Rand area going, and a new find never failed to create a sensation. Not far from the pumping station at the mouth of Goler the Exploration Company of Garlock had opened up a rich vein in their mine. The samples shown in Randsburg and Johannesburg were of quartz with free gold showing all through it.

The year ended with one of the most unusual gold finds ever recorded in the District. The Big Butte had been making news under the new Superintendent-Manager Percy McMahone with 40 men working $40 rock. A milling at the Red Dog converted 24 tons of ore into a gold bar weighing 63 1/2 ounces. At $16 an ounce it was worth $1,016. Excitement rose; a new gold boom could be in the making!

Shortly after the display of the gold bar, there was an even greater surge of interest in the Butte Mine. While McMahone was tracing a break in the formation, he discovered an unusual looking opening. Lights shined into the hole revealed a pile of waste, which was as unexpected as the location of the opening. McMahone had men help lower him into the opening. Moving aside a pile of old newspapers, he discovered 39 sacks of high-grade ore. The cache had evidently been placed there by thieves eight or nine months before, while the mine was temporarily shut down. The hiding place was ideal; the only access was through the small hole on the surface that opened into a supposedly well-filled stope of an old shaft that had yielded only $4 ore and was too poor for further mining. When milled the material in the sacks ran $300 a ton.

Wilkerson family in 1898 at their mine in Summit. Horse in center is hitched to a whim; photographer's rig is in background; and man, far right, is holding a paper to gauge the lighting. *(Author's collection)*

Homesteaders locate their lands near Cantil. *(Author's collection)*

9

From Epoch to Epidemic

Perry Whiting of Whiting and Mead, a long-established Los Angeles firm, wrote of a trip he and a friend made in March 1901. With a hired span of mules and a camping outfit they stopped at Lang's 16 miles east of Saugus the first night, at Elizabeth Lake the second, and in a corral at Mojave the third night. As they described it:

> The wind nearly blowing us across the desert. Next day we drove through Red Rock Canyon along the old wagon trail to Owens Valley. It took 5 days to Indian Wells. (From Los Angeles to Indian Wells–less than a 5-hour trip today.)

They went to examine a mine in the Panamints and on the return trip came by way of Searles Lake, following the 20 mule team route. En route to Randsburg to look at the "wonderful mine,"

> We had to walk beside the wagon with overcoats buttoned up tight to keep warm and endure it all.

Stopping to see the renowned and fabulously rich mining area was the usual thing for Los Angeles and San Francisco travelers if they were in the vicinity. Men who knew nothing, a little, or a great deal about mining had been coming and going for years, and the residents paid little attention to them, being far more concerned with what was happening that affected their own lives.

The building of the big 100-stamp mill at the Yellow Aster was of primary interest as it meant more job security. Each step in the building progress was watched with enthusiasm and bets were laid on the date of completion.

Community disappointment was echoed in the *Randsburg Miner*, Saturday, April 27, 1901:

> The stamps began dropping in the new mill of the Yellow Aster Company, Tuesday morning. The mill

was begun about the middle of November last and the completion of such a building equipped with such heavy and complicated machinery in 5 months shows that the work was energetically pushed from the start.

Had it been generally known, half the population would have been present at this epoch in Randsburg history.

The only display was flags, one from the large pole at the office, the staff at the old mill and one up at the carpenter shop at the tunnel. J.H. Payne and Captain Jones, friends of Mr. Singleton from Los Angeles were the only strangers present at the formal turning of a valve which let on the steam and started the machinery, the latter releasing and dropping the first stamp.

The new mill was built nearly opposite the old mill. Sitting on a high ridge, with a steep descent in both easterly and westerly directions, the mill with its silver-toned galvanized roof and sides could be seen for many miles. It was considered a superb structure; the ore bin in the upper part of the building had a capacity of 1,200 tons of ore, while the stamps, batteries, and other machinery would hold hundreds of tons more in the milling process.

Yellow Aster 100-stamp mill, 1904.
(Courtesy George and Connie Post)

Ore, trammed from the mine tunnels to the rock crusher, could be processed at the rate of half a ton per minute and then be moved by small rail cars into the mill. The round balls in a large revolving amalgam barrel thoroughly pulverized it until every particle of extraneous substance was ground and carried off by water, leaving nothing except quicksilver and gold. The melting room was next door where the large retort and bullion furnace turned out the finished gold bricks.

From the day the mill went into operation until it shut down years later, every man that had any association with the engine and boiler room was proud of that association. It was considered the most complete and modern operation of its kind on the West Coast. Two engines, coupled directly to the line shaft, drove 50 stamps each. The cylinders, lubricated by handsome nickle-plated oil pumps, were driven from the valve gear and needed no attention, since they started with the engine and stopped when the engine came to rest. Nickle-plated high- and low-pressure gages, the painted walls of Greek and arabesque patterns, the highly glossed engines, and side bracket lights were all kept spotless. Two 10-foot chandeliers hanging from the ceiling gave the place an elegant appearance in which the new, time-saving innovations developed by Chief Engineer C. E. Van Meter were artfully concealed.

W. C. "Bill" Wilkerson, who came to Randsburg in 1896, was one of two men handling the amalgamation. Each man was in charge of 50 stamps in the new mill. The operation ran around the clock in 12-hour shifts. Wilkerson told of working the first two years without a Sunday or holiday off, receiving $4.50 for the 12-hour day of considerable responsibility.

The mill shut down only for cleanup or trouble. Wilkerson explained, "We cleaned up four batteries, that was 20 stamps at a time, once a week, which let each stamp run about a month. The take on four batteries that had run a month varied from $10,000 to $20,000. The mill averaged over $100,000 per month on sorted material, but when they started on the big glory hole, it was cheaper to mill everything that came down the chute, though the values per run went down. Later I transferred to the old 30-stamp mill which ran only high-grade; it was a better, quieter job."

Though the big mill was the highlight of early 1901, there were other developments and improvements in the area that the citizens appreciated. The Randsburg Railroad was spruced up with a fresh coat of paint for the passenger cars, the depot, the engine house, and the section house in Johannesburg. An improvement was made at the school, too; fresh water was furnished every day, and the tanks were cleaned daily instead of twice a week.

It was a year of heat and exceptional dryness. Feed was so scarce the stock were dying; cattle stood around Desert Springs and Mesquite Springs where vegetation could not keep up with the demand, but fresh water flowed freely. Desert Springs water fed out through a 4-inch pipe, and Mesquite was running in a full stream down toward the valley floor, crossing where the present Randsburg road follows the foothills of the El Pasos.

Water, a prime necessity anywhere and doubly so in the desert's warmer months, was still inadequate on the Rand side of the valley, affecting both Johannesburg and Randsburg. In contrast, there were artesian wells flowing in the valley to the southwest, the present Cantil area. Garlock's many shallow, hand-dug wells and the deep well at the mouth of Goler, owned and operated by the Yellow Aster Mine, continued to be dependable sources.

Garlock's water was mineralized, but less so than that pumped at the Goler well, which carried 17 different chemicals including lime, zinc, soda, acids, arsenic, and a trace of gold. Although the employees at the Goler pumping station drank that water without any apparent ill effect, the Yellow Aster did not use it in the machinery for fear that the acids would eat out the radiators.

The wells at Garlock, approximately 50 feet deep, had provided the surrounding mining camps with water in their early years of development and were still the main source for many individuals in both Randsburg and Johannesburg, although the haul was long and difficult.

One of the shallowest wells—only 20 feet deep—was owned by Juan Barsarto, an early Garlock resident who was laundryman for the whole area. The boys of the town considered it a privilege to help Juan fill his pails, barrels, and tubs. The water was pulled up out of the well, one pail at a time, each evening and let stand over night to settle the clay that came in the seepage water.

Juan was a favorite with adults and children alike. On his buckboard drawn by two burros he was a familiar sight at the mines and in the towns, where he gathered soiled clothing and delivered it freshly laundered. John Robinson Sr. and his sister Lizzie Robinson Short, as youngsters, would drive down from Randsburg to get meat at their uncle's slaughterhouse and fill barrels of water. John still recalls the delicious tortillas Juan would deliver in Randsburg on order for parties along his laundry route, and the beautiful lace he made for the ladies of the camps.

"I can still see him sitting out in front of his little rock and adobe cabin, crocheting great pieces of lace, so long that they lay in the dirt and were almost black." Robinson added, "One time I asked my mother how Juan could sell such dirty lace to ladies, and her reply was quite revealing to a youngster. All he had to do was wash it, and who had more experience than Juan."

Others, who were youngsters at this same period, told of the delicious tacos shared with Juan. It was considered a great honor to receive an invitation to eat with Juan, to lend him a helping hand, and even get to ride to the edge of town as he started on his route.

Around adults, Juan was considered bashful and would never press an overdue bill. However, he was rarely cheated out of his pay, because others saw to it that the delinquent miner or housewife settled up. That same affection was shown the day two Chinese stepped off the stage at the Doty Hotel.

The usual group of townsmen were lounging around the bar across the street, swapping gold stories and watching passengers alight from the stage. The two Chinese very politely asked whom they could see about a building in which to set up a laundry. One of the enterprising onlookers suggested they have a drink and discuss their needs. With a wink to his friends, he led them all as they followed the Chinese into Miller's bar and accepted the newcomers' offer to buy a round.

With unspoken agreement that no one should trespass on Juan's territory, the local fellows pretended that a new laundry was just what the town needed. They slapped the strangers on the back and expressed a most friendly welcome until the Chinese ran out of money; then they loaded them into the outgoing stage. Men laughed as they told others that night about the would-be laundrymen and visualized their surprise when they roused out of the liquor fog and found themselves back in Mojave, totally broke.

Practical jokes, if they can be called that, were not always confined to the newcomer. In Johannesburg, the town that boasted a city map and a planning committee, there was a gap between actual development and the map. Some residents pushed their property line over into unused lots when the owner was not around to protest, and disputes over property ownership were not unusual.

One of the women in town who already owned a large chunk of property wanted the lot owned by Joe Fossie, a miner. She tried to scare him off with threats of legal seizure and statements that he was on her land. Fossie, a large, quiet man, calmly went about his business ignoring the frustrated woman, but he made one mistake. He went on a week-long prospecting trip, and when he returned, his one-room house was gone! Building and contents had been completely removed.

Fossie was not one to be easily defeated. He started a new house, hauling in discarded railroad ties for the floor and walls and topping the sturdy structure with half of a heavy metal tank for the roof. His only public comment was, "She can't move it or burn it." Whether the vanished house had anything to do with his romance, Johannesburg residents never knew, but shortly after the round-roofed building was completed, he brought home a bride and then went on to build a modern house attached to his monument to determination. Fossie built well, for the structure stood in view from Highway 395 until demolished for the sake of appearances in 1967.

As superior as Johannesburg felt in relation to the unplanned, sprawling Randsburg, it could not hide the wild element, the saloons, and gambling, though their number was far less than Randsburg's. Jack Harrison's saloon next to Teagle's store made up for the lack in numbers; his was a match for any emporium in the desert and carried the reputation of more accidental, self-defense, or no-witness murders than all the rest of the saloons put together in the whole Rand area.

Strangers came and went daily; some came for drinks at Harrison's; others came for supplies, or both. Prospectors came infrequently to load up another grubstake before one more try at the search for gold, and no one paid any attention to them unless there was something unusual about their actions or requests. Thus no one noticed a stocky blond fellow who one day tied his burro to the rail at the side of Teagle's store and went inside to purchase a 20-cent half-gallon bucket.

Teagle's store. Harrison's saloon is at far left.
(Courtesy Mabel Edsel)

There was nothing unusual in his taking the bucket next door to the Harrison saloon and having it filled with tap beer; many miners did that on their way home from work. The bartender filled the pail until foam rose high around the bail and ran down the sides. The stranger carried his beer outside, then braced his feet, tipped his blond head back and took a long drink from the pail. People sitting under the porch shade in front of Teagle's saw him wipe the froth from around his mouth and then reach over to place the pail and the remaining beer over the nose of his burro, remarking, "Here have a long cool one."

The burro pulled its head out of the pail and snorted, blowing foam and beer on the prospector and the onlookers. To everyone's amazement and entertainment the man and his burro shared pail after pail of beer; each time the foam would fly, and the man would cuss a blue streak, belaboring the animal for wasting beer. Some one in the gathering crowd asked the prospector if he wouldn't like something stronger to drink, and the invitation was accepted.

Others bought drinks as the conversation flowed, and some hoped to learn of other good prospecting areas. The fellow gave no name; the ethics of the time prevented anyone from asking; and all they found out was that he and his burro had walked five days from his mine in the Panamint Mountains and had come to Johannesburg for supplies. Usually men talked freely after a few drinks and boasted of the richness of their mine or the size of the nuggets found, but the blond gentleman said nothing of importance, though he carried his share of the conversation.

Recalling that evening, Ed Teagle told of the crowd at the bar and between forty and fifty men at the poker tables. Suddenly the stocky fellow from the Panamints stepped away from the bar and started reciting passages from Shakespeare in a deep, clear voice. A startled silence settled over the room; not a card slapped the table or a glass clanked on the bar. "That fellow gave a performance that needed no stage, no lights, no costume. The characters really came alive. About two or three in the morning he stopped as suddenly as he had started. No offer of drinks could induce him to continue. It had been a memorable evening, and we felt culture had walked right into our little mining community."

Next morning the fellow picked up his supplies, secured the pack on his burro and departed. That night the bar was full again, and the conversation was mainly about the stranger who had put on a professional rendition of Shakespearian plays. Since the performer had given no name, he was soon referred to as Shakespeare. People speculated that he must have been an actor who for some reason preferred the solitude of the rugged Panamints.

As time went by the incident was almost forgotten, when the man reappeared, bought another half-gallon tin bucket, and went into Harrison's for beer. Shakespeare was back in town, and the word passed around. When the prospector was satisfied that his burro had had enough beer, he accepted drinks at the bar. Teagle ordered a Flusey for him, a whiskey three times as large as the present-day shot glass drink, at a cost of 25 cents. Others took turns buying drinks and hoped for another evening's entertainment. People who were not usual patrons of the bars gathered outside, waiting. "When Shakespeare finally moved away from the bar, you could have heard a mouse sneeze; that evening was just as wonderful as the first one."

When Shakespeare left the next morning, no one knew any more about him than they had on his previous trip. Weeks later the same routine of beer for the burro ended again in an evening of drama. Teagle said that Shakespeare's trips for supplies became a regular thing and the community was proud of its actor.

Some people talked of following Shakespeare to find out where his mine was located, but as far as anyone knew, none of the talkers ever tried to travel the miles of sandy washes, scramble over the tufa formation, skirt the dry salt lake, and climb up into the Panamint range to spy on the man who spent freely and gave so much pleasure. The week that Shakespeare was due in for supplies but didn't appear was a time of concern for the whole town. People checked their memories for absences of any of the town's rough characters, fearing that someone might have harmed the prospector. As the time stretched into two weeks, there was talk of a search party, but no one knew where to look once they reached the Panamints. The people were sad and helpless; all they could do was to ask freighters through the area to keep a lookout for any clue to a stocky blond prospector with a wonderful voice and a burro that snorted in his beer.

When an epidemic broke out just after Christmas, concern for the missing Shakespeare was shoved into the background. Smallpox, the dread disease that had filled graveyards of other pioneering communities all across the country had gained a foothold in the Rand area.

Considering the fact that miners, gamblers, merchants, and camp followers moved from one glamorous boom area to another, it speaks well for the healthful nature of the area and the cleanliness of the desert mining camps that until December 1901 no epidemic had struck.

When it came, it was a floater who brought the smallpox. After circulating about the camp, he went to Dr. Baxter, a dentist, to have a troublesome tooth removed. As the dentist leaned over the man he immediately recognized the early signs of smallpox. Dr. Reginald E. McDonald, private physician and doctor to the Yellow Aster Mining Company, was soon made aware of the situation. Before word could get around and start a panic, the doctor enlisted the aid of the Kern County Health Department.

Just after Christmas, approximately 200 people suddenly became ill. No facility existed in Randsburg to handle so many ailing citizens. Where possible, persons were allowed to remain at home under quarantine; others were moved to a huge tent designated as the Pest House. The noninfected stayed a safe distance from the homes or the dreaded Pest House and tried to keep the community running as nearly normally as possible.

Volunteers who were immune due to past illness or exposure or who had no fear of the disease moved about helping the doctor. Some nursed in homes, tents, or dugouts of the miners, and others spent full time at the big tent. As in many another mining camp, much credit was due the women from the red light district. They demonstrated a skill and willingness to help wherever there was a need, whether nursing, preparing meals, cleaning, or sitting the long night hours with the critically ill. Many a "respectable" woman who had looked down on the less virtuous of the town now found tender carc at the hands of the very ones she had shunned.

Before the disease ran its course, over 500 people were afflicted in the communities of Randsburg and Johannesburg. Some were reported ill at isolated mines, and an unknown number of people died.

Yellow Aster Mill ran three shifts around the clock in 1903. *(Author's collection)*

Randsburg Band at the Johannesburg Railway Station as part of a reception committee. *(Author's collection)*

10

Never a Dull Moment

The two towns on the mountain, Randsburg and Johannesburg, were months recovering from the smallpox epidemic and had just settled into the routine of everyday living, mining, gunplay, and labor disputes by the time 1903 rolled around. The new year was destined to go down in the history of the area as an unusually diversified period. It started off with a fire in the upper story of Louie Woodward's two-story dance hall in Randsburg. Butte Avenue was a raging inferno within minutes and most of the buildings again lay in ashes.

Labor trouble was credited with the destruction. A strike had been declared the night before. The mine had shut down, and then the town was almost wiped out. There was little doubt but that the fire was set, for when the volunteer fire company reached their equipment, the hose had been cut, the water turned off, and the rope on the fire bell was missing.

During the labor strike, feelings ran high; even the school children were affected. The union kids sat on one side of the room and the non-union on the other. Some families grew disgusted with the situation and sold out for enough to move them to other employment. But there was one who couldn't leave; all that he owned had gone up in smoke in the fire along Butte Avenue. Tex's One Man Magic Show had been appearing nightly at the Woodward Dance Hall, and when it burned, Tex Lovett's livelihood lay in ashes. Staying to work in the mines to build up funds to equip a new show, he found mining more lucrative than show business and remained to become an old-timer in the Rand area.

Johannesburg had a few bad days early in April when rumor turned to reality; the Randsburg

Railroad had been sold. Until the facts were known, there was concern over the possibility that the line might be abandoned. Santa Fe bought the road that had cost $180,000 for $300,000 and made some immediate changes. The two daily trains were replaced by one mixed daily.

Though Johannesburg was blessed with more water than Randsburg, there still was not enough to fight fires, and when the Pinmore Mill caught fire, the efforts of a bucket line proved fruitless. The mill was a total loss. From then on more and more attention was given the construction of buildings that were more nearly fireproof, and "Dobie" Bowman's skills as a builder of adobe structures were much in demand.

Most of the adobe buildings still standing in the Rand District were built or supervised by Bowman. One of the historically colorful buildings that he completed is the White House Saloon and dance hall that still stands on the main street of Randsburg opposite the post office. The owner of the White House had suffered from the 1898 fires and decided to have a lasting building put up even if it was expensive. Louie Stoll couldn't afford another burnout; for years he had the reputation of running a "tight saloon," allowing very little credit and putting up with no overindulgence that might lead to destruction of furnishings or stock behind the bar.

Like the building, the bar was well built; Yellow Aster carpenters built it of heavy timbers. Although it was not exactly a thing of beauty, it was equal to hard service. An immense dumbwaiter, large enough to hold a man, operated between the card room on the main floor and the restaurant in the partially underground lower level. Its purpose was to supply food to the card players without their having to leave the game, but a dual purpose came to light after the sudden disappearance of a few wanted men.

One of the individuals seen and then not seen in Randsburg was the outlaw Jim McKinney. He was believed to have hidden out in Randsburg after shooting up the town of Porterville and killing a couple of his friends. When McKinney's girlfriend Sarah was seen in Randsburg, it was assumed that she was meeting or visiting him. On April 3, 1903, his presence was definitely suspected as reporters from Los Angeles and San Francisco converged on the area about the same time as a posse arrived from Arizona.

McKinney had committed two murders in Arizona and had headed back to California where he could get help in finding a hideout. His brothers Ed and Jake operated a saloon at Rademacher, about 14 miles north of Randsburg, and Sheriff John Kelly believed that they were concealing the outlaw. Sheriff Kelly knew the area well; he was aware of the endless possibilities for a hideout in isolated mine tunnels and the rough canyons of the El Paso Mountains. He also knew that McKinney and his brothers were just as familiar with the area as he was.

Later evidence pointed to the probability that the outlaw was in Randsburg five days ahead of the posse and had moved on when it became apparent that his brothers were being watched and could give him little help. Ten days later, McKinney met an inglorious end in a gun battle with the law in Bakersfield.

The huge dumbwaiter in the White House Saloon was credited with aiding McKinney in his escape from Randsburg. Others were not so lucky. One outlaw attempting to reach the safety of the waiter found it too far from his place at the bar and was shot before he could get the boxlike vehicle lowered out of gunshot range. Another fellow was enjoying the gold he had stolen from a mine near Fremont Peak when the irate owner, who had seen his teen-age son killed by the thief, rushed in and started firing. There was no time to get to the dumbwaiter so the murderer hid behind the piano. Wires sang as the father's bullets went through the instrument and put an end to the outlaw.

Another interesting feature of the White House was a well-timbered stairway that at one time led from the street to the underground restaurant. At the turn of the stairs, halfway down from the street level, a heavy door concealed the entrance to a tunnel. A hump and seam are still visible in the main street of Randsburg where the tunnel went under the street.

Callahan's Boarding House stood on the site of the present White House prior to the fire of May 1898. Callahan had the tunnel dug when "pay dirt" was discovered at the side of his building. He hired unemployed miners to dig his gold as pay for overdue board bills. Following a small gold vein, the men dug a tunnel right under the street. The vein soon pinched out, but to keep Callahan interested they salted the diggings with higher grade ore from the Yellow Aster and kept their board bill paid until they could get back on the Yellow Aster Mine payroll. Though the tunnel was filled and boarded up in later years, there were many who surmised that it was used for a hiding place in the case of an emergency such as the unexpected appearance of the law.

Excitement was not confined to the adult world. Fred Osborn, a first grader in 1904, never has forgotten his teacher, Mrs. Higgins. "She was very strict; one time when I happened to study the wrong spelling lesson, she took me to the front of the room and struck my hands with a ruler. One of my friends, Ivan Blackwell, laughed, so she took him to the washroom and washed his mouth out with soap. While the teacher was out of the room, Ivan's sister ran home and told her mother what was happening to Ivan. Next thing we knew Mrs. Blackwell came into the room with a blacksnake whip and chased Mrs. Higgins around and around the room, snapping the whip at her. Finally the teacher escaped from the room and ran home. That ended school for that day."

Osborn, in his notes on memories of Randsburg, stated that the wild burros wandering around the area added excitement to a boy's life, too. "The Houser boy and I would catch one and try to ride it. Many are the times I got bucked off into a pile of tin cans that littered the yards around some of the shacks. When the burros wouldn't cooperate, we'd ride a coaster wagon down the hill and then hitch a ride back up to Randsburg by fastening our wagon behind a water wagon coming up from the valley with water."

Giving a boy's-eye view of the Yellow Aster operation, Osborn described a tour he and his mother were given by the company's assayer, who boarded in the Osborn home. The big mill was noisy with the long row of heavy iron stamps pounding up and down, and the vibration was so great that it seemed difficult to keep one's balance. In another room a stream of water mixed with quicksilver was bringing the ore down a sluice box. When young Osborn sighted a good-sized nugget, the assayer told him to pick it up. The gold continuously slipped from his grasp, and finally the man explained that it was due to the quicksilver.

In the furnace room they watched a worker take a pair of tongs and pull out a small steel barrow and pour liquid gold into a mold that stood nearby on the floor. Steam filled the room as water was poured over the hot metal; when cooled the gold brick was taken from the mold, pounded with a special hammer, and then moved to the assayer's office. The assayer drilled a hole in the brick and weighed the drillings, explaining that this was the way he found the value of the brick; young Osborn wondered why he didn't just weigh the whole brick. The last they saw of the gold was when the company superintendent came with a horse and buggy to take it to the Wells Fargo office down on the main street of Randsburg.

First load of Yellow Aster ore ready to be hauled 10 miles down the mountainside to the Garlock mill.
(Author's collection)

At the time Osborn was visiting the mine, the *Los Angeles Times* was publishing "Men of Achievement in the Great Southwest." The description of the Yellow Aster differed greatly from what had interested the boy.

> Kern County has the largest gold mine in the Southwest and for that matter, one of the largest on the Pacific Coast, the Yellow Aster in the Rand District. The Yellow Aster Company is a closed corporation, and is owned and controlled by John Singleton, C.A. Burcham and Dr. Rose L. Burcham. The property is in every respect one of the most

perfectly equipped on the Pacific coast, having two stamp mills, one of 30 stamps, the other of 100 stamps, its own water plant, which includes pumping station and several miles of pipe line, affording an ample supply for all purposes. The mines are electrically lighted, every possible improvement and convenience to facilitate work and keep down expenses being provided. It has a good record of production and a future that is rich with promise of increasing production for years to come.

Probably of more interest to the local citizens was the *Randsburg Miner*, weekly newspaper that noted in the January 16, 1904, issue that:

> Marguerite Roberts was called into Justice Court, charged with keeping a house of ill fame. Judge Manning continued the case.

A very small item on the back page mentioned that the post office had been moved to an adobe building next to Marguerite Roberts' dance hall. Two weeks later the *Miner* headline read, "Marguerite Roberts Dive Investigated."

J. R. Rush had filed a complaint with the County Board of Supervisors on the basis that Marguerite Roberts' business was "not fit for licensing" and in no uncertain terms added:

> 1. Runs a dance hall in the very center of the business section of our town, so situated that we and our children are compelled to pass and repass this abominable brothel every time we go to the postoffice, drug store, meat market or Wells Fargo Office.
>
> 2. We are continually being insulted by the inmates of this brothel, who are common prostitutes.
>
> 3. Therefore we ask in the name of high heaven and common decency that your honorable body will revoke the saloon license.

Within the month Marguerite was hauled into court for allowing a minor, Marguerite Fredricks, age 15, a drink in her establishment. Witness in the case was a minor, Helen Fredricks, who in turn had been seen drinking in Fay LeRoy's saloon. As the case proceeded it became clear that the two girls were sisters, Carrie and Margaret Buffington of Oxnard, and that each had been paid to spy on the rival establishment. Each saloon was fined $25 and warned against any such antics in the future.

Meanwhile mining went on as usual; mines closed, opened, went under new management, or were sold. The long-worked Sunshine was running $150-a-ton ore, the Butte milling of 67 tons at the Red Dog cleared $5,300, the Kenyon's leasors were taking out $200-a-ton ore, the Reward shut down, and in April of 1904 the Yellow Aster was in legal trouble again.

The Yellow Aster was being sued for $10,000 for the death of Martin Bush, killed at the Goler Well pumping station March 29, 1902. Twenty-eight-year-old Bush, working for $100 a month as engineer in charge of hoist, had been the sole support of his parents. Their claim was based on the fact that Bush had been with the Yellow Aster only a short time and was not acquainted with the dangers of the pumping system. He and two others had been ordered into the well to make repairs; when they gave the signal to be pulled up, the engine failed. After some delay the bucket holding the three men was brought up to the surface, but Bush was dead from asphyxiation.

Six months later, up at the main mill in Randsburg, Engineer F. C. Frickie was scalded over half his body when a steampipe burst. He had been back to work only a short time when he was sent to the Goler pumping station as night foreman, a job that was less demanding and would give him an opportunity to recuperate. As he was routinely throwing a piece of waste saturated with oil into the fire box to ignite the oil, Frickie's face and arms were badly burned by an explosion. The oil pipes had been turned so low by the day shift that the fire had gone out before quitting time; no one had been aware of the condition that allowed oil to go on dripping into the fire box unburned. The hot steam pipes turned the oil to gas that exploded when Frickie went to light the oil for the night operation. The explosion burst the firebox and furnace door, throwing burning oil all over the man, who had been knocked to the floor by the blast. Frickie, blinded by the oil, still managed to gain his feet and shut off the oil to the main boiler before collapsing.

Crippling accidents were an accepted part of mining, often leaving those involved with no means of livelihood. Feeling was voiced that a big company like the Yellow Aster could do more for

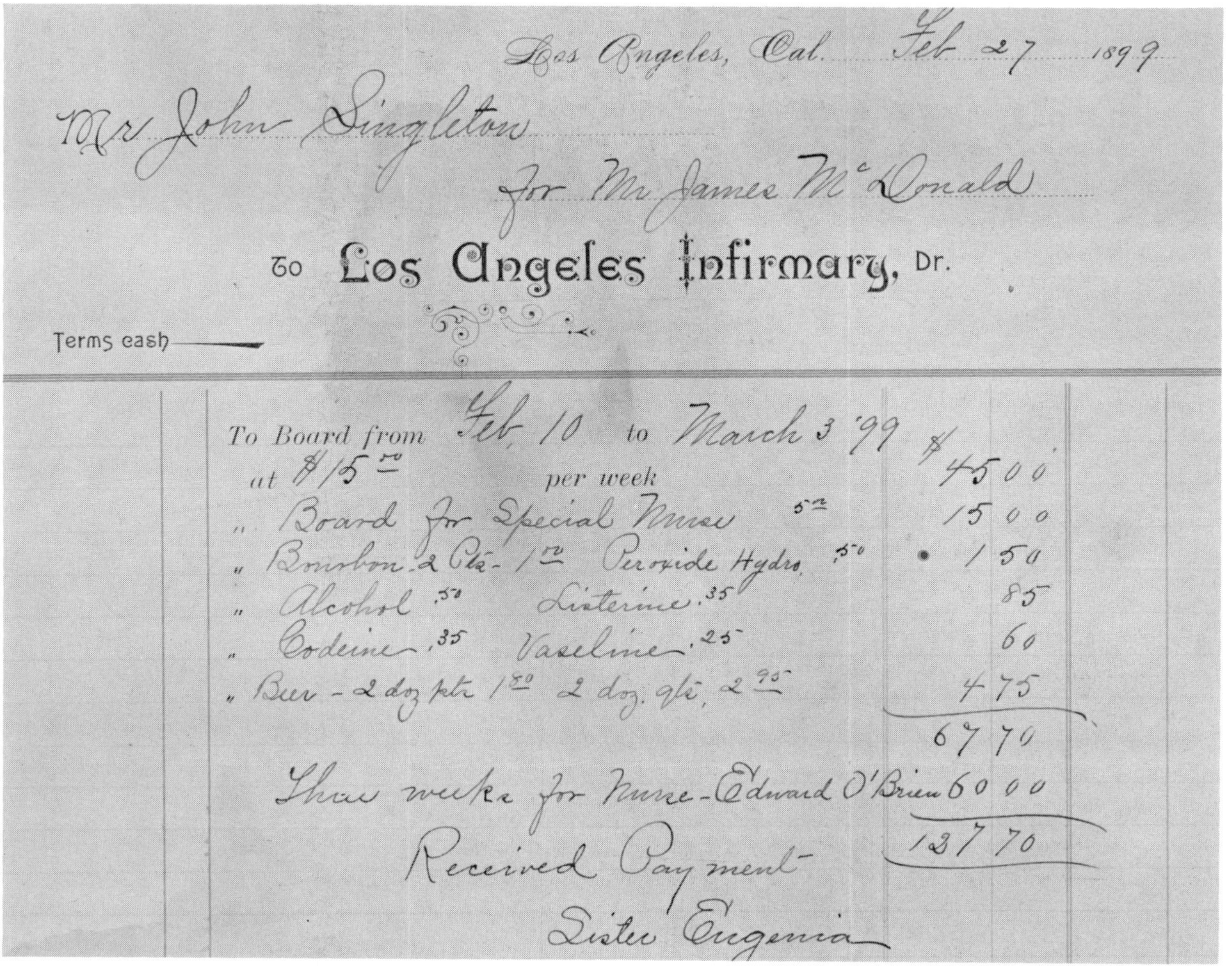

Los Angeles, Cal. Feb 27 1899

Mr John Singleton

For Mr James McDonald

To Los Angeles Infirmary, Dr.

Terms cash

To Board from Feb 10 to March 3 '99 at $15.00 per week	$45 00
" Board for Special Nurse 5.00	15 00
" Bourbon 2 Pts - 1.00 Peroxide Hydro. .50	1 50
" Alcohol .50 Listerine .35	85
" Codeine .35 Vaseline .25	60
" Beer - 2 doz pts 1.80 2 doz. qts, 2.95	4 75
	67 70
Three weeks for Nurse - Edward O'Brien	60 00
	127 70

Received Payment

Sister Eugenia

Nursing care, bourbon, and beer are among the remedies listed on this bill paid by John Singleton for a mine employee.

those injured in their employment. But the owners of the mine may not have cared or did not hear about accidents because their lives were involved elsewhere.

John Singleton, President of the Yellow Aster was seldom seen, and then would just be in and out claiming poor health. Occasionally word concerning him would reach the Rand area, such as the time he was a refugee from the San Francisco earthquake.

The local paper carried a reprint news item that John Singleton, Los Angeles millionaire, his wife, and her sister had been staying at the Palace Hotel when the quake struck. They gathered their clothes and got out immediately, hiring an express wagon for $25 to take them to the Casino near Golden Gate Park. The following day and at enormous cost they got back to the Palace. They had paid $1 apiece for eggs and $1 for a loaf of bread, and with the change they had left had bought a little ham. The millionaire had been stranded without a cent until finally someone recognized him.

C. Austin Burcham, Vice President of the Yellow Aster was away so much of the time looking after other mining interests and activities in San Francisco that W. H. Storms, Consulting Engineer was advanced to Assistant Supervisor to relieve Burcham of some of his duties.

Though the Yellow Aster apparently had a good supply of water being pushed up the hill from their pumping station, the communities of Randsburg and Johannesburg still suffered from the lack of even enough water for household use. When the much-publicized Marguerite Roberts moved a shack to the rear of her establishment on lower Butte Avenue, the local paper carried the following on its front page:

> Conditions are fast ripening in that section for a repetition of the big fire last June. The business men who own the adjoining buildings, not having recovered from the losses occasioned by the last fire are not in a position to erect adobes.
>
> A fire is liable to start during a drunken orgy any night, and once started would quickly wipe out the whole lower half of town, businesses, homes, rooming houses, hotel, and on these windy nights it is hard to tell where it would stop.
>
> Under the law our officers have the power to enter all places of unsavory reputation frequented by women and arrest and punish as vagrants everyone found therein.

In February 1904, a month after Marguerite created an additional fire hazard by moving into a shack, M. R. Irwin, Superintendent of the Randsburg Water Company announced the sinking of two shafts to provide more water and the installation of two reservoirs holding 100,000 and 30,000 gallons, respectively. The new development sounded like the solution to a long time problem. However, a few months later when school boys saw a fire coming from a mattress lying against Tom Smith's house, there still was no water available nearby and they had to beat and stomp the fire out.

The following year there was not enough water to prevent fire from completely destroying a Santa Fe boxcar at the Johannesburg depot. The car was loaded with merchandise for Ballarat and fireworks for the Fourth of July. A car coupled to the burning one was loaded with dynamite. E. F. Wolever, station agent, and others uncoupled and pushed the car with explosives away as fireworks were shooting off in all directions. It was thought that a leaky or broken carboy of nitric acid caused the fire.

A short time later the Cozy Hotel of Johannesburg caught fire from a gasoline stove in the kitchen, and the entire building was in flames by the time a passerby saw it and gave the alarm. It was early Sunday morning and the guests escaped with what clothing they could grab. Only one guest, Fred Bell, was burned on the face and arms as he tried to save $40 in gold and a tin can filled with nickles. A bucket brigade carrying water from all over town was credited with saving buildings near the hotel and possibly saving the entire town.

That November Ed Teagle's house caught fire from a heater in the bedroom. The Yellow Aster Mine whistle gave the alarm, and though there were plenty of men to fight the fire, there was no water to aid them, and the structure burned to the ground. Warnings were posted all over town, the volunteer fire fighters practiced their skills, and the *Miner* carried articles in almost every issue in hopes of keeping every one aware and careful of fires.

The May 25, 1905, issue of the *Miner* hit at another sore subject in Randsburg: "It is a regrettable fact that we have people who have no more pride than to use Butte Avenue for a swill barrel." The editorial went to some length in describing the "stench of reeking filth," composed of "dishwater, rotten potatoes, vegetables, refuse, etc."

In spite of the "reeking filth," there was the social life of Butte Avenue like the "elegant supper" and charivari of the newlyweds, Minnie Williams and Harry Green of Illingsworth and Green Company, at their home on Butte. And there was the Yellow Aster Social Club replacing the Randsburg Social Club, which had died of dissension. The new club grabbed the lease on the Kennedy Building and left the old club without a home and then proceeded to welcome everyone into membership. They became an active social influence in the community by sponsoring a dress affair in the form of a Saint Patrick's Day ball, sponsored the ladies star basketball team, and put on an oyster supper, followed by a masquerade ball.

In between the dances there were home talent shows organized by a fellow who had been in the theater. For those not wishing to appear on the stage, there were evenings of dancing. Local musicians furnished the music, and since there was always a shortage of women, the young girls had an opportunity to learn to dance at an early age.

The Junior League held basket socials; each woman brought a basket of food and ate supper with the man who bid the most for her basket. Entertainment was furnished by League members. Program notes listed such numbers as

Song and March by 10 little girls, "Touch-not, Taste-not."
Recitation, "Which General," by Merle Van Pelt
Recitation, "When Willie Goes to School," by Geraldine Houser
Solo, "A Baby For Sale," by Imagene Lidy

Other activity on Butte Avenue included meetings of the AOUW Desert Lodge No. 326 in their building on Butte every Thursday evening. Degree of Honor, Desert Queen No. 64 met every Tuesday in their hall on Butte. The Independent Order of Foresters met in the AOUW hall on Wednesday nights. Special entertainment coming to town generally used the Union Hall as did Professor A. L. Athans' Comedians, appearing for five nights. Admittance price: adults 25 cents, children 15 cents.

In addition to the social life of Butte Avenue there were summer picnics and watermelon feeds in the evening when the temperature had cooled down. Lanterns and sometimes a campfire would light the faces of the participants as they ate, listened to a favorite story teller, or sang the latest songs.

Each of the mining communities had baseball teams, and Sunday afternoons found most of the citizens, regardless of age, watching and cheering their team. Competition ran high, and betting was part of the game's popularity. The first ballgame in the area was played in 1897 between Randsburg and Garlock, with a score of 11 to 5 in favor of Garlock. Teagle Brothers Store in Johannesburg later sponsored a ball team that won fame for that community by beating every team that would play them. Ed Teagle once asked Los Angeles to play them but was turned down. "They would have been disgraced to have been beat by a little mining town team."

A big rodeo was scheduled at Koehn Springs for May 14, 1904, with 25 vaqueros from Kernville, Lancaster, Glenville, and Tehachapi putting on an exciting exhibition. Riding, roping, bulldozing, and branding skills kept the large crowd cheering and betting. A side interest that day was the flying machine the "Mayor of Kane Springs," Charlie Koehn, had almost perfected. People teased him about setting up an air route between his way station and Randsburg, but seriously promised a big celebration if he did fly into town.

Shortly after the rodeo, high winds damaged the incomplete airship, tore down a large barn in Garlock, blew over the Houser livery stable and wagon sheds, and ripped fronts off houses in Randsburg. Charlie's air route to Randsburg had a setback from which it never recovered.

After a hot, quiet summer, the fall of 1904 brought a number of visible changes. Houser Brothers moved several houses from Garlock and Johannesburg to Randsburg. Porter Brothers, Johannesburg freighters, were busy transporting mining machinery to Ballarat and the mines in the Panamints.

Suddenly Randsburg became a community aware of its responsibilities. Two youths tried to solve problems that should have had adult help at an earlier stage, but the townspeople had shown little concern. Willie Pitzel, 13, and Thomas Ritter, 15, walked from Randsburg to Kane Springs, then stole one of Charlie Koehn's horses when they could walk no farther. The boys rode the horse to Mojave, turned it into a pasture, and started walking again; their destination was Bakersfield.

When questioned by a man who gave them a lift on his wagon, they revealed that Ritter's father was dead and his mother was seriously ill in the county hospital in Bakersfield. When his mother was removed to the hospital, the boy had been left alone with no funds. The only way he knew to better the situation was to get to Bakersfield where he could at least be near his mother. Pitzel went along on the trip to escape the mistreatment meted out by his drunken, widowed father. The boy had been living on what money he could scrape together by running errands about town.

Randsburg made Thanksgiving a special day that year, providing aid for the convalescing Mrs. Ritter and finding employment for her son Thomas. The town's men saw to it that Pitzel paid for living expenses of young Willie, who now had a home and an understanding mother in Mrs. Ritter.

Randsburg women put on a bazaar.
(Courtesy Emma Beher

No. 198

Yellow Aster Mining and Milling Co.

Randsburg, Cal. FEB 20 1900 190

Pay to W. A. ... or order $50.00

Fifty ... Dollars

To The National Bank of California,
Los Angeles, Cal.

John Singleton PRESIDENT.

SECRETARY.

Yellow Aster check was appropriately printed on gold-colored paper.

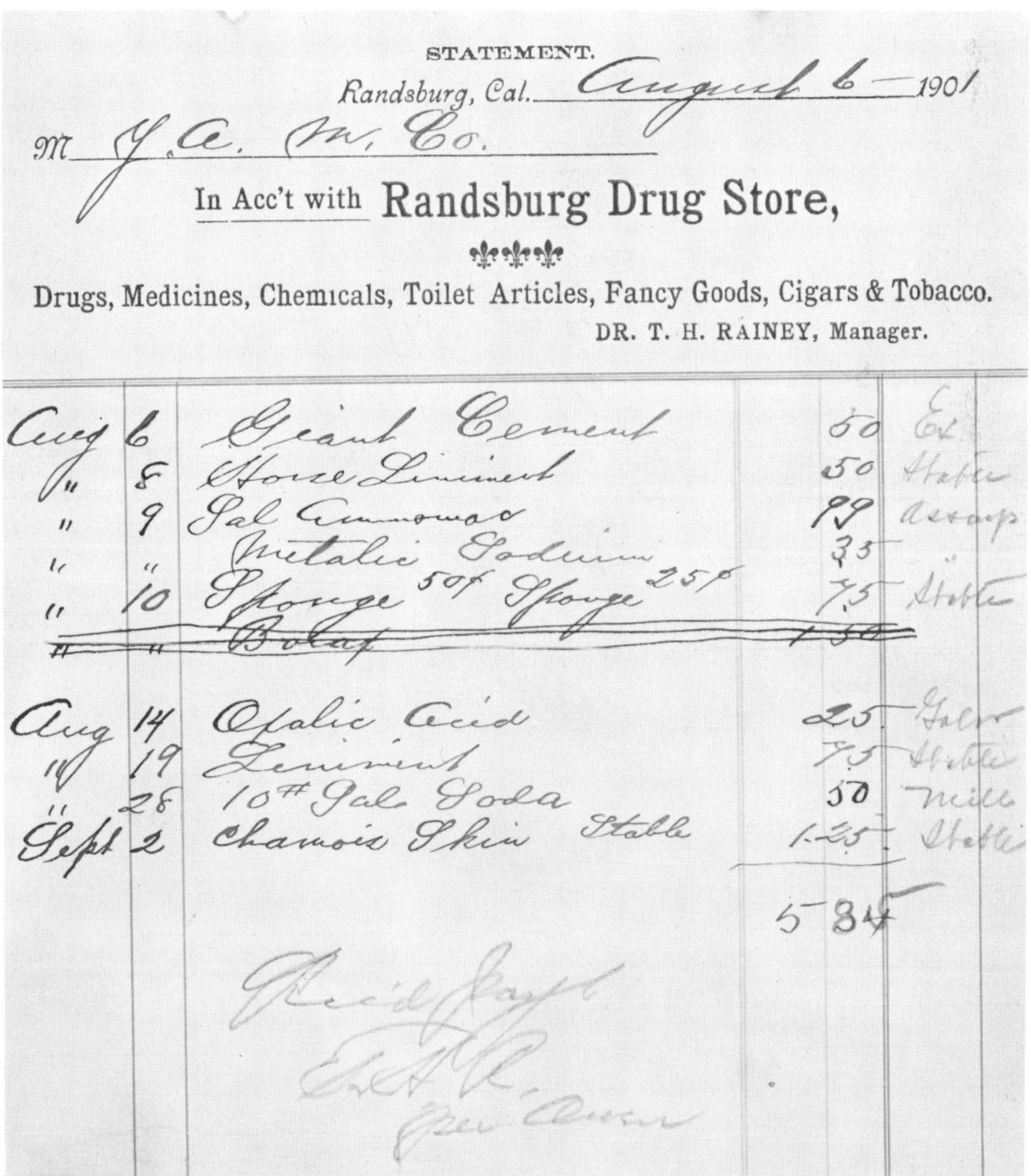
STATEMENT.

Randsburg, Cal. August 6 1901

M Y. A. M. Co.

In Acc't with Randsburg Drug Store,

Drugs, Medicines, Chemicals, Toilet Articles, Fancy Goods, Cigars & Tobacco.

DR. T. H. RAINEY, Manager.

Aug	6	Giant Cement	50	Ex
"	8	Horse Liniment	50	Stable
"	9	Sal Ammoniac	99	Assay
"	"	Metalic Sodium	35	"
"	10	Sponge 50¢ Sponge 25¢	75	Stable
~~"~~	~~"~~	~~Borax~~	~~1.50~~	
Aug	14	Oxalic Acid	25	Galv
"	19	Liniment	75	Stable
"	28	10# Sal Soda	50	Mill
Sept	2	Chamois Skin Stable	1.25	Stable
			5.84	

Statement of Yellow Aster Mine's account carefully annotated as to the use of the items.

Garlock, 1898. Doty two-story hotel, seventh building on the right; Lillard Hotel across the street; school at end of street on the left. *(Author's collection)*

Will Barnum with load of sage and greasewood fuel for steam boilers of Garlock Mill, 1896. Sam and Charles Barnum's feed lot in background. *(Courtesy J. H. Martin)*

Boarding house run by Mrs. Archie Martin in Goler Canyon, 1894. From left: Tom Lackey; Will Barnum; Earnest Huntley; two unidentified boarders; Dave Van Buran, a prospector seated; two unidentified boarders; Arthur Martin; J. H. Martin leaning against father, Archie Martin, seated by post; Charles Barnum standing; Walter Martin seated; McClusky, a prospector; Mrs. Archie (Mary) Martin, mother of the Barnum and Martin boys; unidentified woman; Frank Barnhan; and Pat Letty at edge of picture, behind canteens.

(Courtesy J. H. Martin)

Sunshine Mine east of Randsburg.

(Author's collection)

Flag raising day, Yellow Aster Mine, Randsburg, 1896. F. M. Mooers, original discoverer is fourth from the left in white suit and straw hat.

(From collection of his grandson, Douglas F. Mooers)

Flag raising day at the Yellow Aster Mine, 1896.
(Douglas F. Mooers collection)

Yellow Aster Mine and Mill, 1901. *(Author's collection)*

St. Elmo Mine, 1897. Mojave stage on right.
(Courtesy Kern County Museum)

Dust of the long line teams and freight wagons could be seen along the valley floor as they approached Garlock, 1893-1906. *(Author's collection)*

A new stage line celebrates the first run.
(Los Angeles County Museum collection)

Randsburg snowstorm, January 12, 1913. Pete Osdick and Dora ? having snowball fight back of Catholic Church, left center. *(Courtesy G. Schmidtmann)*

Dr. Rose Burcham's house in Randsburg.
(Author's collection)

Interior of Lillard Hotel, 1898. From left: Zeek, "Tom," "Dick," and Mrs. Z. T. (Nancy Jane) Lillard, who built the business with her cooking. *(Author's collection)*

S. Masters, horse shoeing, general hauling, carriage works, Randsburg, 1898. *(Courtesy Mrs. Coren Church)*

One of the long teams hauling oil to the Goler pumping station, 1903. *(Author's collection)*

Everything from the glory hole was hauled in and milled. Yellow Aster Mine, 1901.
(Courtesy Kern County Museum, Randsburg)

John Singleton's mansion in Los Angeles.
(Courtesy George and Constance Post)

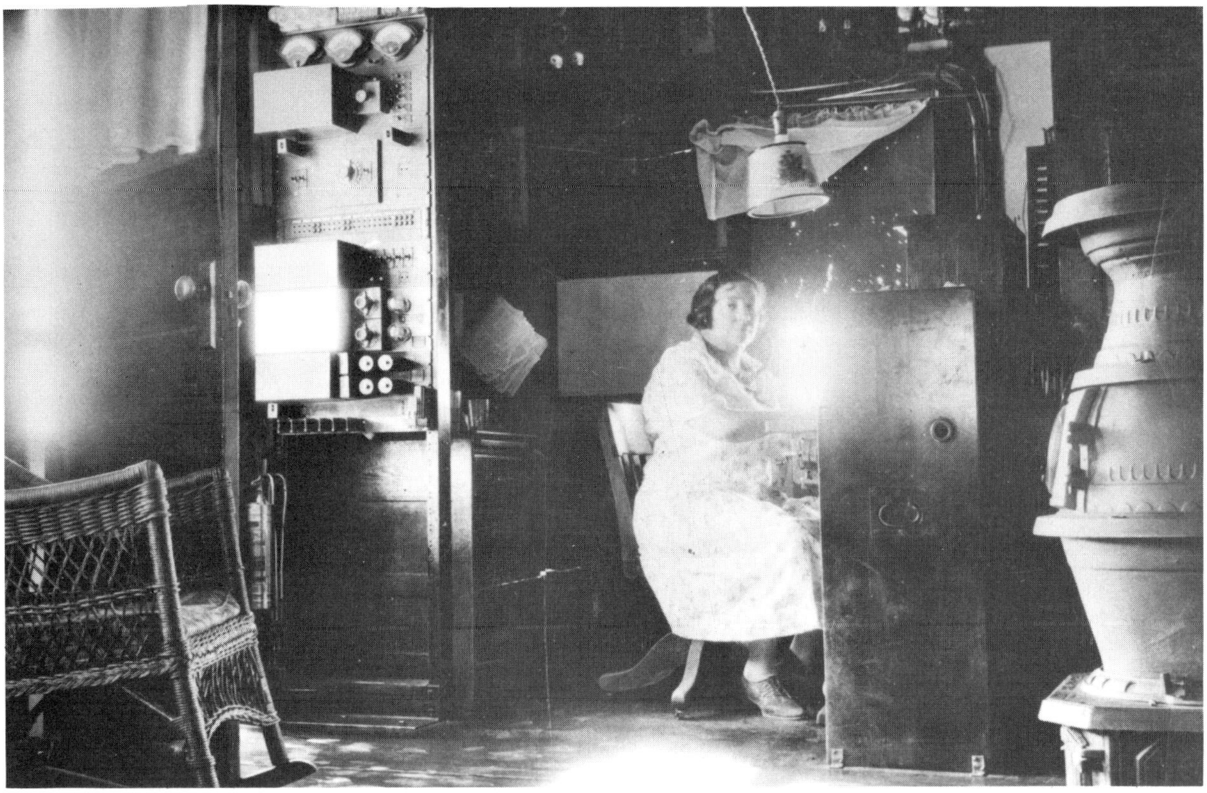

Elizabeth Hirschier Suchey, relief operator in Randsburg telephone office, 1933. Equipment to left is part of the long distance relay system just installed.

(Courtesy Marie H. Johnson)

Doty Hotel in Garlock, July 4, 1897. A. J. and Sarah Doty standing third and fourth to right of door; little girl, granddaughter of the Dotys, Laura Meeker Cole, seated in front of her mother, Laura Doty Meeker; Grace Gibne, teacher, top of ladder to right. Streetlight at corner was Garlock's modern touch.

(Courtesy Laura M. Cole)

Archie Martin home in Garlock (ruins still standing), 1896. From left: Mrs. Archie (Mary) Martin, Arthur Martin, George Barnum, J. H. Martin, Harry Martin, and Charles Barnum. *(Courtesy J. H. Martin)*

Hotel Johannesburg, 1898.
(Los Angeles County Museum collection)

Sierra Ruez, left, and Jim Gabriel. Yellow Aster Mill, 1912. *(Author's collection)*

Investors look the land over as their guide scans the map. *(Author's collection)*

11

Not All Was Gold

In those days near the turn of the century, picturesque, gold-embossed shares were sold on any likely prospect, whether mines, railroads, oil wells, or water companies. As a chance to share the profits of a growing nation by grubstaking, the new development through shares caught the fancy of many. To own one-millionth part of a lucrative business seemed a good investment for a dollar.

The Rand mining area was offering products other than gold for the investor. Excitement ran high when coal was discovered five miles north of Garlock at the foot of Black Mountain and development was started by the Pacific Coast Coal Company. More interest was created with the news that the Southern Pacific Railroad had contracted for coal to be delivered to Mojave for use in their engines. When, later, coal gradually dropped out of the conversation and other interests took over, few realized that coal had been delivered to the railroad, the engines had been stoked, but had failed to move. The coal burned but did not generate enough heat to build steam to move the engine.

Approximately five miles to the east of Garlock an oil rig was being hauled up into Goler Canyon by the Houser Brothers freight line. Asphalt had been discovered, and it was believed to be a sure sign of oil. Tools, lumber, and tents were deposited just above The Narrows, and drilling was started with hopes of a new boom for the canyon that had been crawling with gold seekers just 12 years before. But a heavy flow of water was struck at 16 feet, and the project was abandoned, along with the high hopes of investors.

However, failure in any venture was quickly forgotten in the glow of other projects. Prosperity was assured; the Baltic Mine sold for $75,000 the same week the oil drilling stopped, and rumors were flying that something big was developing in the tungsten field near St. Elmo.

When tungsten was first discovered, there had been some publicity and the welcome mat was put out for miners and investors. But later, a railroad spur was added at St. Elmo, and the public was no longer welcome. The 15 miners employed at the mine suddenly were absent from the local saloons. The air of secrecy was not merely imagined. E. H. Moyle of the Engineering and Equipment Company, Los Angeles, was refused admittance to the management's tent. Moyle, a frequent visitor at the camp, had previously been given tours of the mining operation and had been presented with numerous ore samples and been asked for advice. When Moyle visited in February 1906, Superintendent C. Taylor would only say that the company order was: "Keep everyone out and everything in."

Ore was sorted in a sealed tent, and within a month seven more miners were added to the payroll. The new railroad spur made possible the loading of cars right at the mine. Shipments were tagged for the Barstow Mill and Liverpool, England. The mine area that had been called Tungsten or St. Elmo now was officially named Atolia.

Less notice was taken of a mill being built four miles south of Garlock. Jack Watson, a Los Angeles builder, was hired by a group of investors to erect a salt mill on the edge of Koehn dry lake. When the floor of the mill was completed, Watson staged a dance and invited everyone for miles around. During the dance he began to notice the young woman sent out to sew salt sacks. They soon married; Watson later remarked that the marriage had quickly failed. "She couldn't cook or keep house. All she could really do was sew salt sacks!"

At that time the process of claiming salt from the dry lake depended entirely on the weather. Good rainfall followed by hot sunny days added up to salt. The whole operation hinged on there being enough rainfall to drain from the Rand and El Paso Mountains to fill the lake. A heavy brine was formed from the minerals carried in by the runoff and the high concentration of salts in the lake. Following the rains there was a long period of waiting for evaporation to form salt crystals along the rim of the lake. The first cakes of salt were cut by hand, but this process was too slow and was soon replaced by a gasoline-driven saw on wheels. Men turned each square foot block of salt on end and scraped a half to a full inch of mud and gypsum from the bottom before loading it into ore cars headed for the mill. An assembly line of men handled the whole job from the cutting of the salt chunks, through the milling process, to shoveling the end product into hand-sewn cloth sacks.

Watson, like the investors in Los Angeles, thought the salt harvest prospects were so good that he bought 1,000 shares at $1 each. But the rains were inadequate that year and production was so low that the company went broke. When the Diamond Salt Company bought the mill in 1911, Watson sold his 1,000 shares for the total sum of $15. Real production started in 1914 by the Consolidated Salt Company on the eastern end of the lake and the Fremont Salt Company on the western end. In 1928 both companies came under the ownership of the Long Beach Salt Company, present owners.

A few miles to the west in Red Rock Canyon Rudolph Von Hagen was having water "shares" printed that never reached the public who might have profited from the water distribution plan. Von Hagen had come to the area in the late 1890s and had bought or taken up claims on all the land he could get his hands on in the Red Rock country. An issue of the *Inyo Independent* in 1893 stated that Von Hagen had washed $150 worth of gold in four days; it was local belief that he took out thousands of dollars in gold in a very few years. His store, stable, and feed yard, named Ricardo for his son, were money makers, too, as they were a much patronized freighting and stage stop.

Von Hagen expressed the belief that there were great undiscovered pockets of gold in Red Rock Canyon, and although he continued to both dry and wet wash for gold production, he decided on a faster process. He would sell shares to develop a water system that would channel Red Rock's stream to homesteaders and ranchers in the valley to the south. The water would be sold, and Von Hagen would scrape the gold accumulation off the bottom of the channel.

A letter dated July 1, 1906, signed by E. Newman, North American Society of Civil Engineers, Keeler, California, contained the engineer's findings and recommendations after examining Red Rock for Von Hagen's proposed water development:

> The canyon should be tapped where the water accumulates in the cove, and a tunnel bored through at this point will be safer than a timbered tunnel in the wash and will cost less money. The approximate fall of the canyon is 4 ft. to the 100, or 4%. The length of tunnel and mouth of tunnel must be determined by soundings to ascertain depth to bedrock in cove and by leveling down the canyon until a point is reached where the top of the ground is on the same level as bedrock in the bottom of canyon.
>
> The tunnel must be timbered and tight lagging used on the sides and top until rimrock is reached. From this point the tunnel will be in solid rock and will need no timbering.
>
> I would suggest that you build a tunnel 4 ft. wide, 6 ft. high. It is impossible for me to tell you how much water you can develop and the length of the tunnel. I think you will have to run a tunnel 1,000 to 1,200 ft. long. Regard to cost of tunnel, I think I could run it for $10.00 per ft., but you may not be able to contract for the work at $20.00 a ft.

The shares were already printed but never presented to any prospective buyers. It would seem that the engineer's figures caused Von Hagen to drop the plan.

Rudolph Von Hagen, 1906. *(Courtesy Martin Engel)*

Though Von Hagen's tunnel never got under way, another one in the area did develop. To the northeast, high in the El Paso Mountains, William H. Schmidt was starting to drive a 5- by 7-foot tunnel through a mountain, a distance of over 2,000 feet.

Schmidt, like his three sisters and three brothers, was ill with consumption when he left Rhode Island for California. It was 1895 when 24-year-old Schmidt arrived in the far West, the last hope for regaining his health. Employment with the Kern County Land Company kept him outdoors, and the climate was so beneficial that by 1900 he was able to spend part time as a prospector, climbing mountains, and camping out in rough country.

With the Rand mining area booming, gold fever replaced consumptive fever, and Schmidt located a number of claims on and around Black Mountain in the El Paso range, northwest of Garlock.

For the next 32 years Schmidt worked summers on ranches along the South Fork of the Kern River to finance his mining operations in the winter. In 1906 he started work on a tunnel that took him until 1938 to finish. Over half of his 83 years were spent working with a four-pound single jackhammer, and dynamite, and thousands of hours with crowbar, pick, and shovel.

Schmidt was not a man to discuss his private business, but for the patient, not too inquisitive, he would get around to explaining that the tunnel through the mountain would save hauling ore over miles of rough, steep, and sandy roads to reach a shipping point. From the end of the tunnel it would be a short drop to either a main road or a rail siding. During the digging of the tunnel he sold off good mineral claims, and by the time of its completion he had little need for it.

The tunnel, a monument to one man's determination, is a tourist attraction today. Few visitors realize that it was driven through a semi-broken down granite that shattered when blasted, that Schmidt picked and hammered the loosened rock, and loaded the material into an ore car that he had pulled into the tunnel by hand and later pushed back out, loaded for the dump. For economy's sake he sharpened every piece of steel that he used, mined by candlelight, cut the dynamite fuzes dangerously short, and used tools until they were worn down beyond handling.

In the early days Schmidt was known as Jackass Schmidt, because he always drove a burro or two hitched to a light wagon when he went into Randsburg for supplies. The name, not meant as derogatory, simply included the common term for the burro. In recent years the tourists and "Johnny-come-latelys," repelled by the term, changed it to Burro Schmidt, which meant the same but sounded more refined.

As the fragile, stooped man grew older, he kept more to himself and in his late 70s stopped driving to town. Fred Weimholt, who worked for the Black Mountain Mining Company in Last Chance Canyon would pick Schmidt up for a weekly trip to town where the old miner would buy few items other than boxes of soda crackers and a gallon of wine. Weimholt later remarked that Schmidt's yard was immaculate; no tin cans or bottles cluttered the place; and he would warn strangers, "Don't throw your 'ceegareets' in my yard."

While Von Hagen dug Red Rock Canyon for gold and Schmidt dug a tunnel through the mountain, Sam Cuddeback and other cowboys riding for Kern River's cattle king, William Landers, were pulling Landers' cattle out of the swampland around Kane Springs. It took 15 to 20 cowboys five days to brand the calves rounded up in the area now known as Cantil. Cuddeback, who stayed at the Munsey ranch while working cattle, on his way out to check on a herd one morning saw fresh tracks of two people and a horse that came nearly to the Munsey gate, then turned and retraced their trail. His curiosity aroused, the young cowboy followed the tracks and caught up with the party three miles to the south. A man and woman were walking beside a horse that was carrying a small girl. From their movements, Cuddeback felt sure they were in trouble, but when he asked if he could help, the man just mumbled, "No . . . no."

After riding back to the ranch, Cuddeback told Mrs. Munsey what he had seen and his belief that they were lost and exhausted, but were afraid to tell him. Equal to any situation, Mrs. Munsey put a big pot of coffee on to boil while she fixed food and had Cuddeback saddle her gray mare. She rode out ahead with Cuddeback coming along at a slower pace in the buckboard, with the coffee and food.

By the time he caught up, Mrs. Munsey had the group stopped and had found that they were desperately in need of water. Assured of her friendship, they drank and ate while telling of their attempted trip from Los Angeles with a ton and a half of household goods on a big beer wagon pulled by the one horse. The weight of the wagon alone was too much for one horse, and attempting to pull the load in October heat had nearly killed the animal. They knew nothing of the country and a number of times had been within a half mile of water, which the horse would have found if they had known enough to let him lead the way.

The Munsey boys went after the beer wagon and brought it to the ranch while the strangers rested and decided what to do. In a few days enough money was raised for fare back to Los Angeles by selling the wagon, its contents, and the horse, and by donations from neighbors. Then the family took the stage out from Cinco, the nearest stage pickup point.

Cuddeback knew well the discomfort of thirst. In July 1907 he and a number of other men were moving some of Landers' cattle from Tehachapi to Kane Springs. "Everyone had celebrated the Fourth in Tehachapi, and some of the boys were still carrying a pretty good load. Ben Roberts was in charge and he didn't stand for any monkey business; everyone was expected to do his part, regardless of hangovers. Citano, the cook, was driving the wagon with the food and two barrels of water; Roberts had told him exactly where to set up noon camp, just ahead of the drive."

As the day grew hot, one cowboy after another reported to Roberts that he was out of water and dying of thirst, which was no doubt true considering the previous night's celebration. The cattle were moving along well, and as they neared the midday stop, Roberts told one fellow after another to go ahead and catch up with the cook wagon and water.

"When we reached the designated point," Cuddeback recalled, "there was no cook wagon, no water, no cowboys. Those that had stayed with the herd were really suffering, and Roberts let them go on to locate the wagon. Finally there was only Roberts, Big Johnnie Weldon, and myself left with the cattle. After hours in the heat, eating dust kicked up by the cattle and trying to stretch a half day's supply of water into a full day, we, too, were suffering."

As they passed within a half mile of the old deserted 16-mile house, south of the present Cinco, Weldon rode down to see if there was a drop of water anywhere. "He came back with an old oyster can filled with alkali water. It wasn't good but it was wet."

Cuddeback added, "The cattle seemed to know they were headed for water at Kane Springs, and the old ones strung out, with the young ones falling into line too. They would probably have reached the spring whether there had been a driver or not, but we stuck with their slow pace and it was 2 a.m. the next morning when we reached the sleeping camp. The cattle had a good drink and so did we; then Roberts kicked the cook out of bed and told him to start walking for Landers' ranch in Kelso Valley and see how it felt to be without water when the sun was high."

Many lives were affected by water. Jobs at the Yellow Aster Mine in Randsburg depended on a supply being boosted up from Goler, a job that involved 18 to 20 employees in the pumping operation alone. Scott Seals, hired to oversee the pumping, brought his family to Goler in 1905. Mrs. Seals cooked for 17 of the employees stationed there. The boarding house was about a city block from the pumping station. Sleeping quarters were provided in a bunkhouse on the opposite hillside.

Mrs. Seals enjoyed a good joke and frequently played pranks on her boarders. One April Fool's Day she baked cupcakes and put a bit of cotton in some, seeing that a big fellow called "Doughbelly" got the one really loaded with cotton. The other fellows were all good sports, never letting on that they had run into anything unusual, just placing the unchewable addition on their plate. When the big fellow took a bite and couldn't get his teeth through the cupcake, he put the whole thing in his mouth. After chewing for some time, he still had a mouthful. When he was offered another cupcake, he jumped up from the table and shouted, to the delight of all the other men, "Mrs. Seals you're trying to kill us all."

Since Doughbelly was noted for loading his plate before anyone else had their portion and then going back for seconds, saying, "If no one else wants this I'll just whoop it all out," without giving anyone else an opportunity, the men were delighted when anything slowed him down.

Another boarder told Mrs. Seals that beans nearly killed him and that she would have to provide other food for him when she served beans. After the subject was forgotten, she served a vegetable loaf and that particular man ate a good portion of it; for the next few days she kept inquiring about his health until he asked why she was concerned, adding that he had never felt better in his life.

Mrs. Seals was pleased with his remark and explained, "That was bean loaf you ate so much of the other day and I've been worried that it would kill you." A few days later he told Seals that he was quitting; when asked the reason for his sudden departure, he explained, "You can't trust the food you get here."

Mrs. Seals was also known for her relentlessness when it came to all-night poker games. Though she enjoyed cards as much as the men, there was a time when a game should end if men were to work the following day. If they failed to heed her warning and kept on playing, she would upend the table, scatter cards and money all over the place and send the men scurrying for the bunkhouse with a tongue lashing that left no doubt that she meant what she said.

Managing horses with the skill of a man, Mrs. Seals would drive any animal to Randsburg for groceries even though she was warned against their running away. Though she demanded obedience of the teams, she did not abuse or stand for abuse of any of the stock at the pumping station. One day when a tenderfoot brought the long team and oil tank down from the railhead at Johannesburg, he tried to force the team the wrong way around the circle drive used by the oil wagons. The lead animals knew where they were supposed to go and ignored the driver, who became so angered that he started beating their heads with a hammer.

Mrs. Seals looked out her kitchen window and saw what was happening; rushing out with the broom in her hand she beat the driver until he got away from the team. Upon returning to Johannesburg the driver asked for some other type of work as he didn't wish to haul any more oil down where there was a crazy woman. Burcham drove down to Goler to see what the trouble was, and as soon as he heard what the driver had done, the fellow was fired. Burcham, once a rancher, would keep no one on the Yellow Aster payroll that abused a horse or mule belonging to the company.

Seals was not only in charge of the pumping operation, he was also skilled at making cable. Great rolls of wire were hauled in, and he spent 15 months hand-braiding a new cable for hoisting water out of the well. When they were putting the cable into use, the fellow running reel in the hoist house wasn't paying attention to the cable and nearly wound Seals onto the reel. He was nearly to the top of the framework within a few inches of the reel when another workman heard his cries and rushed into the hoist house, shoved the operator aside, and stopped the cable.

Such mishaps always served to reemphasize that each man's life depended upon the other fellow's doing his job; there was no job in the area open to a man after he had endangered others working with him.

The sound of a gunshot was not strange to the Rand mining camps from the very beginning, but by 1905 the population considered itself more civilized and shootings were hushed up. Nevertheless, in spite of the highbrow attitude, a gunfight still stirred up considerable interest and betting on the outcome. When the trouble involved a quarrel between O. E. Cheesebrough and John Whitmore, friends and leasors on the Magpie, it created quite a ripple of excitement; they were the last men one would expect to have disagreed over anything. The trouble first became known when Whitmore went to Doc MacDonald to get a slug removed from his right arm, and Doc demanded an explanation.

Later, in the courtroom the facts brought out were based on each man's statement, since no one else was around at the time of the shooting. The men had quarreled over the mine; Cheesebrough claimed to have gone down the ladder to the 100-foot level to get his tools and Whitmore had followed. Both men were armed. Cheesebrough said Whitmore raised his gun; Whitmore claimed he didn't move his pistol and had only followed Cheesebrough in hopes of talking him into reconciliation. Charges were dropped, and the case was dismissed.

"Case dismissed," was a frequent solution to a shooting, especially when Jack Harrison, saloon owner in Johannesburg, was involved. In January 1907 Harrison was charged with the murder of J. H. Arnold. Among the witnesses a Dr. Hall testified that "complications plus the gunshot were cause of death." Others stated that the boys were all having a good time, that there had been no ill will at the party, and that the shot, fired in the dark, was accidental and without malice. Harrison was released.

That same year Jack Harrison, Shorty Warren, and "Kitty" were on trial for the murder of Henry T. Wright. Harrison and Kitty were absolved, and the decision handed down was that Warren had shot in self-defense: case dismissed.

Bev Robinson, a young fellow freighting for Teagle's store next door to Harrison's saloon saw a bullet dig a hole in the loading platform at his feet. (The hole is still visible today.) A youth just in from Ballarat was waiting around in front of Harrison's for the next outgoing train and had refused to join others in a friendly drink at the bar. The guys hanging out at Harrison's figured the kid had gold on him and was heading home. A shot rang out, there was a lot of movement, and the body was robbed right in daylight in front of the saloon. A court hearing was held with no defense for the murdered miner. The boys in the vicinity of the shooting were local citizens; the decision was that they had shot in self-defense, and the matter was dropped.

The local *Miner* gave very little space to the shootings at Harrison's, but on one occasion a front page item explained much. It was the editor's reply to a complaint that not all the news was printed:

> If we published all that happened we would be with the angels. In order to please the people we must print only the nice things. All the news is all right when it is about the other fellow.

Johannesburg Machine and Blacksmith Shop, 1912-14.
Celebrator unknown. *(Courtesy Mabel Edsel)*

Twenty-mule team crossing Fremont Valley is heading northeast to pick up ore in the Panamints. Team is hauling hay to deposit at way stations along the route. *(Author's collection)*

12

Freighting and Railroading

Much has been recorded about the mines, but little about the freighters who made mining possible. Much of the economy of the Rand area depended on the long-line drivers and the teams that moved supplies and equipment in from the railhead at Johannesburg or from Mojave.

Fred Conoway was one of the men with headquarters in Johannesburg, whose life was freighting. Conoway's parents moved to the Rand area in 1905, and his father drove 8- and 10-horse team wagons for the H. D. & L. D. Porter Freighting Company of Johannesburg. The family lived at Old Wells, east of the present Red Mountain. In order to get an education, young Conoway walked 10 miles to school. Every other day he would stop on the way to operate the pumps for Charles Chase, owner of the water system, and would miss part of that school day.

The pay he received for tending the pumps was collected by his father; when Fred became old enough to help with the teams, there was even less school and still not a cent for his own use. At the age of 14 he left home and took the job of driving stage and mail between Johannesburg and Ballarat. Most trips were routine, but there were some incidents that used all of the young driver's ingenuity. On one trip he removed the seats from the stage and placed a cot across the full length of the vehicle in an effort to get an ill man out to the doctor. The man had blood poisoning in one leg, and on every grade Conoway and the man's wife would turn the man around so the infected leg would be up-grade to keep the blood out of the injured area; at the top of the hill they would go through the same process again for the downhill trip.

A trip out from Johannesburg with a load of city slickers was good for laughs as Conoway later described it. Before leaving town he had put a 2-gallon canteen in at their feet and warned them not to waste it as it was some miles to the next water point. They kidded him about trying to scare them, saying that if others traveled this route, there had to be plenty of water. A few miles out of town when the temperature was climbing, he warned them again about drinking too much of the water, knowing that they could become ill from overdrinking. He was not too much concerned about a water shortage since he carried extra water on top of the stage.

The city fellows gave the young driver a bad time, making snide remarks and calling up to him for answers to questions so that they could belittle him more. They laughed when the stage hit a bump and the canteen flew out; Conoway would have stopped to pick up the water container under other circumstances, but this time he drove on. Soon one of the fellows asked for water and was informed that all there was lay in the road a few miles back. As time went on, they begged for a drink, but the driver told them they had to wait until the next water stop. He didn't add that it was only a few miles ahead. "By the time we reached the next water stop, they had about talked themselves into death by thirst. They were good guys the rest of the trip."

Conoway hauled gold out of Skidoo every other week when his stage run was extended beyond Ballarat. The shipment would be $10,000 to $40,000 in gold. He would put it under his feet with the mail sacks and drive as if he had nothing more than passengers and mail. He remarked that, "There was never a holdup and it's probably due to the fact that I didn't take any extra precautions that would tip off anyone to there being gold on the stage."

Even with thousands of dollars under his feet, Conoway said that as the heat climbed and the straight stretches grew monotonous, he would doze. The horses knew the way, and it wasn't until they hit rough country that he would have to be awake and alert.

What Conoway didn't tell was related by Ed Teagle, "That boy was a born teamster, and no job was too big for him to tackle. Only time I ever saw him down was when his wagon hit a boulder and threw him off and about 7 tons of freight ran over his foot. Conoway's helper got him back up on the wagon and on into Johannesburg, where he insisted on being left at his cabin and didn't want anyone around. The helper went for Dr. Rose Burcham, but Conoway wouldn't let her or anyone else in, holding them off with a gun." At the time he was hauling for Teagle so Ed told him that the least he could let them do was get him some liquor to help ease the pain.

The liquor was accepted, and as soon as it started to take effect, Teagle and some of the biggest men in town held him down while a hot poultice was put on the foot. "Every time he got moving around we would pour some more whiskey down him, even poured some on his foot. When the train was ready to pull out, we loaded him into the baggage car and got him out to the hospital. That foot was as flat as your hand, but he didn't lose it and before long was back on his wagon."

The Randsburg Railroad had brought supplies and mining machinery closer to the Panamint and Death Valley country, but materials going north were still being hauled from Mojave. By June 1908 a new line was on its way out of Mojave toward the east that in time would connect with the Carson and Colorado from the north. Men coming out from Los Angeles to work on the new line had to wait in Mojave until night for transportation out as far as the rails had been laid.

Chair cars carried the workmen; box cars and flats transported materials for construction of both the railroad and the Los Angeles aqueduct. Passengers were a mixture of Slavic, Italian, Mexican, Irish, German, and American laborers. Chinese cooks and Japanese flunkies kept to themselves at one end of the car. Gaudily dressed women of all nationalities circulated from car to car, leaving a trail of heavy-scented perfume that mingled with the odors of stale liquor, food carried in baskets, and unbathed bodies.

The train would hardly be out of the Mojave station before arguments would start between the laborers, or some liberal drunk would go down the aisle forcing a drink on each passenger. "You tipped up the bottle and swallowed, pretending to take a drink or run the chance of getting your head bashed in for being unfriendly," remarked one of those early train riders.

A few miles out of town the train would slow to a stop, puffing patiently in the dark while conductor and brakeman put off the drunks, troublemakers, and free riders who had no pass or work order. Little except loading facilities were at the Chaffee, Cambio, Trescope, and Pine stops, but Cinco was a large camp with men and materials spread out between the foothills and the new rails.

Great piles of construction materials stood in boxcars on a spur track. Acres of desert were covered with wagons, scrapers, and fresnos, along with stockpiles of hay, sacked grain, coal, pipe of all sizes, and boxes of dynamite. Near the foothills was the tent city with the smaller shelters for the men and huge tents for mess hall, store, and blacksmith shop. A rough board structure provided hospital facilities, with a nurse or two and a cook on resident duty.

Cinco was a noisy, active place, with men calling back and forth as they loaded wagons. There were drivers yelling orders to their long teams as they wove in and out of the piles of pipe and strung out over the miles of sand trails into the camps beyond the rails. Metal clattered and screeched as it rubbed together and finally settled into place in the big wagons.

A crew of 400 men with 900 horses and mules had quickly pushed the rails from Mojave to Cinco, but from there on the problems of crossing Red Rock's wide drainage channels, sand, and steep grades slowed the work. The main line detoured around Red Rock Canyon because of the yearly threat of floods and the expense of cutting down steep grades.

Because of the need to get supplies up into the canyon for the construction of the aqueduct, the Red Rock Railroad was organized. The branch line was built from the Cantil siding, northwest for 9 miles to the Dove Springs aqueduct site. But the Red Rock Railroad had a short life of 22 months, and that was filled with misfortune. One flash flood laid waste the greater part of the line; smaller floods continued to harass the little railroad until it finally was given up as a poor investment and was dismantled in 1910. Only parts of the old roadbed and a turnaround remain to tell of its existence.

In the meantime the main Southern Pacific line had reached Garlock and had gone on to El Paso summit and then northeast, where it faced a real challenge. A shoofly (temporary track) curved up and over the summit; helper engines pushed the train over the steep grade until a tunnel could be built near what became Searles Station. The track was so constructed that at the top of the grade the helper engines could turn around. The old shoofly roadbed is still visible from the main line, a steep climb that warranted the building of onc of Southern Pacific's longest tunnels, 4,340 feet, completed in 1909.

Running along the foothills of the El Paso Mountains the mixed train left Mojave at 6 a.m. daily, reaching Cantil at 7:22 a.m., Ceneda at 7:45, Garlock at 7:57, Goler at 8:12, Rand siding at 8:26, Teagle siding at 8:40, and Searles at 8:55 a.m. Smoke streamed back over the cars that rocked, jolted, and lurched at what 1908-10 passengers considered a breakneck speed.

One of the last steam locomotives to take on water at Cantil. *(Courtesy Loren Burch)*

Water was taken on at each stop. While the fireman filled the iron monster, active persons crawled off the train and impatiently walked up and down the loading platform. Others sat fretting in the sweltering heat of summer or freezing drafts of winter as winds circulated through the coaches. Cinders covered clothing and irritated eyes and were present whether the windows were open or closed.

Drinking water came from a cooler at the rear of the car. A granite dipper was used by everyone until the luxury of paper cups was installed. One fellow en route to Trona with a group of other young men, later told of the water cooler being iced before they left Mojave. They took advantage of the ice when the conductor wasn't looking, by dropping their bottles of beer into the cooler. "It was difficult to wait for the beer to cool, especially when passengers kept going for a drink of water. Finally we all gathered around the cooler to shut out others until the beer was repossessed. The conductor unexpectedly returned before we had it all out, and raised a real uproar!"

A boy that frequently rode the train between Garlock and Inyokern looked forward to passing through the old Summit diggings east of Teagle and watching the heads pop out of the dugouts in the dirt wall facing the railroad. Miners and their families would wave to the passing train and then disappear back into tunnel or dugout.

As the rails reached farther north, the type of passenger changed from laborers and camp followers to families, railroad and mine officials, gamblers, and the gay, pretty girls of the dance halls. Most were headed for the money spots where their money was already invested or where they hoped to hit it rich, one way or another. There were those who would work hard for every cent they got while others would "work" at getting it away from them.

The ways of making a living were about as varied as were the travelers, and Conductor Bill was not beyond adding to his own wages. There are those who remember seeing him collect fares just as on any other line, but as the train neared Mojave he would put the money in his hat and toss the contents into the air. The money that caught on the bell cord would go to the railroad, but that which fell on the floor was pocketed.

The coming of the Southern Pacific line to the valley north of Johannesburg was a jolt to the Randsburg Railroad, now owned by Santa Fe. The competition was not only felt in the freight business, but passenger traffic fell off too. In an all-out effort to hold the passenger traffic, Santa Fe put on a triweekly Pullman service from Johannesburg to Barstow, but many of the Rand area residents found it less time-consuming to take the stage down to Garlock Station, catch the Southern Pacific to Mojave, and then go on into Los Angeles or north to San Francisco.

Passengers on the Randsburg Railroad were plagued with the same cinders and discomforts of the red plush seats with horse hair padding that worked through the upholstery to be more scratch than plush. Fewer stories have survived concerning travel on that line, but one old-timer recalled being part of a youthful group that thought it would be a Halloween prank to board the train that evening without tickets or money and have a joy ride. When the conductor was informed of their trick, he seemed amused and let the riders enjoy themselves for a couple of miles. The train came to an unscheduled stop, and suddenly the train crew converged on the group; before they realized what had happened, they were standing in the desert with the train steaming off into the distance. It was a long, dusty, and dark walk back to town; for years none of them saw any humor in the turn of events, and little further mention was made of the incident.

Train smoke on the horizon was a signal to the youngsters of the town and anyone else free to go to the station that the daily peak of interest was approaching. Jimmy Bertino, a small boy when his family moved to Atolia in 1910, was one of the preschool boys who delivered the daily paper to his parents. At the warning whistle far out of town, the youngsters, coming from every direction, would jostle for best positions along the tracks. As the train puffed to a stop, the conductor standing on the lower step of one of the passenger cars would toss a rolled newspaper to each waiting boy. No paper was missed; tucked inside was a small chocolate candy bar to assure safe delivery to the boy's home.

In many ways life in Atolia was as unique as the paper delivery method. Young Bertino and his friends were fascinated by the first car to come into town, a Stutz that had a remarkable number of movable knobs and screws. In the absence of the owner, the boys freely examined the novelty and discovered that by removing a cap from a stem sticking out of the wheel, the wind whistled as it rushed out. The sound was captivating so they let down all four tires while they listened to the wind. The owner had to pump up the tires by hand, and parents administered spankings, but the next time the Stutz was left unattended, the temptation was so great that the experiment was repeated.

The deep, dangerous ditches all around Atolia were a playground for the boys as they battled imaginary Indians, outlaws, or claim jumpers. For a bit of spending money they searched the ditches and level areas for tungsten spuds, or ran errands between the mill buildings. There was always something to do for excitement.

Girls had their excitement too. It was the winter of 1908 and snowing when Pearl and Ruby Langdon and their mother got off the train in Johannesburg and "took the beat-up-looking stage, pulled by two horses," for Randsburg. As the stage pulled into the main street, youngsters ran along, hung onto the side of the stage, and shouted, "Two new kids in town! Two new kids in town!"

The new arrivals stayed at the Houser Hotel until their stepfather located a two-room house. It was as cold as the hotel, and the girls had to stay in bed for the next two days to keep warm. Finally they braved the snow-covered, slippery streets and went to the post office, where the town kids started pelting them with ice balls. One hit Ruby in the forehead, knocking her down. She quickly recovered and grabbed the boy that had thrown it and knocked him into the snow. From then on the two new kids were accepted and had no more tormenters.

For these city-raised girls the mine camp was a fascinating, wonderful place where everything was different from what they had previously known. There was the horse-drawn stage instead of a street car; fresh meat came in one day a week and was sold at the stationary store as well as the grocery store; and when anyone shouted "fire," everyone in school would rush out, the boys taking a shortcut through the windows and the girls streaming out the door. "It was a kind of unwritten law that everyone got outside of all buildings when there was a fire, for whatever was involved usually ended up in a heap of ashes. There was only a hose cart with a little hose; sometimes the man with the white horses would rush up and hitch to the hose cart to get it to the fire."

Considering all the hazards around the mine camps remarkably few children were involved in accidents, but occasionally they did get themselves into difficulties. Once, the Munson boy and another small boy about the same age decided to go down into a mine; neither wanted to go first, but finally the Munson boy got up courage and started down the ladder. Then, clinging to the top rung of the ladder he decided that he wanted no part of it and was going to return to the surface. His friend objected to the change in plans and kicked the Munson boy's fingers loose from the ladder, sending him down the 90-foot shaft. Scared and crying, the other boy ran home and told his mother that the Munson boy had fallen down a mine. Rescuers brought the unconscious child to the surface, badly battered and with both legs and arms broken. He was one of the few ever to survive a fall down a mine shaft.

White House Saloon bar built by Yellow Aster carpenters in 1903. This photo taken in 1946 or 1947 shows from the left: Bruce "Curley" Carroll, unidentified man, "Garbage" Bill (or "Rags" Bill) and his two dogs, Grover Kane, and Cecil Wise, with Fred Gray reflected in the mirror. Bartenders: Steve Campbell and George Hadley.

(Author's collection)

13

An Undesirable Element

Most of the day-in-and-day-out patrons of the numerous saloons in the communities of the Rand district were lonesome, disillusioned, or restless individuals. They clung to the companionship of others like themselves, lounged in the security of things they were going to do, and worked only so that they could remain in the barroom brotherhood. The more disillusioned no longer tried, and became freeloaders trading admiration and a listening ear for drinks.

Family men, active miners, and the dust-coated, bewhiskered prospectors just in from the desert were the periodic patrons who enjoyed the atmosphere of leisure and fellowship but were not addicted to the clink of glass, the snap of cards, or the coaxing laugh of a siren of the night moving in the dull, smoke-hazed light of oil lamps.

In Randsburg alone 10 to 15, or half, of the saloons and gambling houses were open any hour of the day or night. An early resident remarked, "We never had enough good miners, and the no-goods seemed to flock in from every direction. The honest citizens were so busy making a living that little attention was given the questionable element that was increasing in the saloons and lounging about the streets." It was kind of a joke when one of them told a local boy to pay for the drinks; when he refused, the boy was warned that if one of the Dirty Dozen asked for something, he had better get it or there would be trouble. When asked to buy drinks the next time, he would have to buy or he wouldn't live to refuse again.

Randsburg suddenly became aware of a gang that was shoving people around and that proudly claimed membership in the Dirty Dozen. Constables changed frequently, and it was becoming increasingly difficult to find replacements. Obviously the Constable needed more community backing; a meeting of the Citizen's Committee was called. Their first action was to post handwritten notices all over the camp to the effect: "The Citizens of Randsburg have organized to enforce the laws. Ten Deputy Constables have been appointed and any riotous and threatening conduct will be suppressed and punished."

M. J. "Tex" Lovett with daughters Clementine (Elton) standing at back, Elsie, left, and Catherine (Christenson), right. *(Courtesy Clementine Elton)*

The "Horrible Swede," head of the Dirty Dozen, laughed at the threat, and the gang continued as if nothing had changed. Guilty or not, they were now credited with every robbery, fire, and shooting that took place anywhere in the area. As things got worse the Yellow Aster Mining and Milling Company asked M. J. "Tex" Lovett to clean up the place no matter whose toes got stepped on. They would keep him on the mine payroll, since the little pay received for the Constable job wouldn't keep his family. Lovett agreed to accept the offer if he could have the authority to discharge anyone from the Yellow Aster Mine that in his opinion was an undesirable.

Both adults and children felt the tense, unhappy atmosphere. Men shipped their families out to Los Angeles until conditions improved. Others, like Frank "Rip" Van Pelt, did the next best thing; as electrician at the Yellow Aster Mine he could only get away on Sunday, but he made it a point to get his young son Merle out of town Saturday night and all day Sunday. They hunted dove and quail, camping in long unused cowboy shacks, or visited diggings where Van Pelt knew the miners.

In spite of his father's efforts, young Van Pelt witnessed activities that his parents would rather he had not seen. The time he and other small boys were playing in the Van Pelt yard and heard a fellow shouting from a second-story window left a lasting impression. The man was shouting at a woman walking along the street below his window; the woman turned and shouted back. Right then as the boys stared, the man lifted a gun to his head, pulled the trigger, and toppled out of the window, dead.

Merle Van Pelt in later years remarked, "I delivered the weekly newspaper and hurried to get around town before dark, leaving the main street until last as there were streetlights on one side. I'd do the dark side as fast as I could 'way up past the Catholic and Episcopal churches and finish down the lighted side at a more leisurely pace. I never let anyone know why I delivered papers on the run."

However, from the day Tex put on the badge, conditions started to improve. He hunted down every member of the Dirty Dozen and a number of other troublemakers. Young Judge E. B. Maginnis backed up Tex with the ultimatum: "Get out of town or there will be a hanging."

According to one account, the night after Judge Maginnis voiced the get-out-of-town decree, Lou Cuddeback had all her chickens stolen. The following morning the chicken heads and feet were found hanging on the front door of the Judge's house. A wild bunch lived in a tent down below the Cuddebacks, and the young lady figured that is where her chickens had gone, so she went down and accused the tent dwellers of the theft. As she stood in the tent door while they strongly denied knowing anything about her chickens, Lou noticed a hump under a blanket on one of the cots. It was not the shape of a sleeping man. Before they could stop her, she rushed in and yanked the blanket off a stack of dressed fryers.

Gradually, as one after another of the rough characters left town, the people relaxed, and Tex began to have the easiest time since taking over the job of cleaning up the town. But the peaceful state wasn't to last long; a rash of fires kept breaking out. Some thought the Dirty Dozen were hiding out nearby and taking this means of getting even with the community for driving them out. Fire was as great or a greater danger to the population than had been the Dirty Dozen. Tex and others believed it the work of an amateur, possibly a firebug among the respected citizens. When the source of the fires was finally discovered, it turned out to be a gang of young boys with a nine-year-old leader.

"That little guy was a tough one," Lovett recalled. Tex caught him stealing after a few of the other boys had already been rounded up. "The only way I got information from him was by telling him that the other boys said he would not steal anything smaller than quarts of beer or whiskey. This made him mad; he said he stole just as many little bottles as the rest and then named the members of his gang." Tex never did get over the boy's ordering a glass of whiskey and a porterhouse steak in a restaurant where they stopped to eat en route to juvenile authorities in Bakersfield.

Tex was noted for his tact in situations that arose during his years in office. When intertown rivalry at baseball games would flare into fights among the rooters, Tex would decide he was needed elsewhere and depart with the remark, "This ain't no place for me; I don't want to ruin anyone's fun."

Probably one of the most dangerous periods in the history of the Rand mining area occurred during World War I. Sixty or more paid thugs were sent into Atolia to sabotage the mines. Tungsten, a necessary mineral in the process of making armaments, was Atolia's offering to the war effort, and damaged mines would mean the end of production. Realizing the importance of its mines, the town kept a watchful eye for anything or anyone that might create trouble. When a suspicious number of strangers began to seep into the area, 35 miners were deputized. Between the local men and Federal agents most of the gang were jailed before any damage was done.

The variety of nationalities working the mines, mills, and construction jobs was a natural situation for quarrels and gunplay. Nationals tended to group together, fight together, and if enough moved in, would form their own settlement. Usually the trouble was between different nationalities, but at one time the law officers and the Citizens Committee were aware of a strong undercurrent in the Mexican colony that sat off by itself between Randsburg and Johannesburg.

Hands off was the word as long as they handled the situation within their own group. When trouble spilled over right on Butte Avenue, however, the situation could no longer be ignored. It was payday with the usual spree; the night was December 21, 1912. A group of Mexicans were grouped on one side of the street, talking happily in their own language until two shots rang out. They all disappeared into the dark except Salvador Alvarez, who lay dead in a pool of blood in front of the Degree of Honor Hall. Everyone in the Mexican settlement was questioned, but because they did not understand English or had not been in town that night, the law came up with no answers. The following day Estabon Aguilar was intercepted as he boarded the train for Los Angeles. He admitted the shooting but claimed it was in self-defense. When he verified his story by showing a bullet hole through his leg, the matter was dropped.

Few of the troublemakers ever returned to the area after they were run out. But Big Scott, considered one of the meanest men known to the district, did return after spending years in prison for a killing in Goler Canyon during its gold rush. Scott returned to his former diggings in Goler and asked merchant Charles Richards to grubstake him until he could get gold out. Scott came into town with a good poke of gold one day and started drinking; the merchant wanted his pay, but Scott put him off. The big fellow soon left, supposedly for another bar; when the merchant left for home, there was one shot and he was found face down with a bullet in his back.

After that no one saw Scott for more than a year. When suddenly he returned, drinking and boasting of his good gold claims, a drunken brawl developed. Someone hit him over the head with a billiard cue. The death certificate read: "Died of pneumonia."

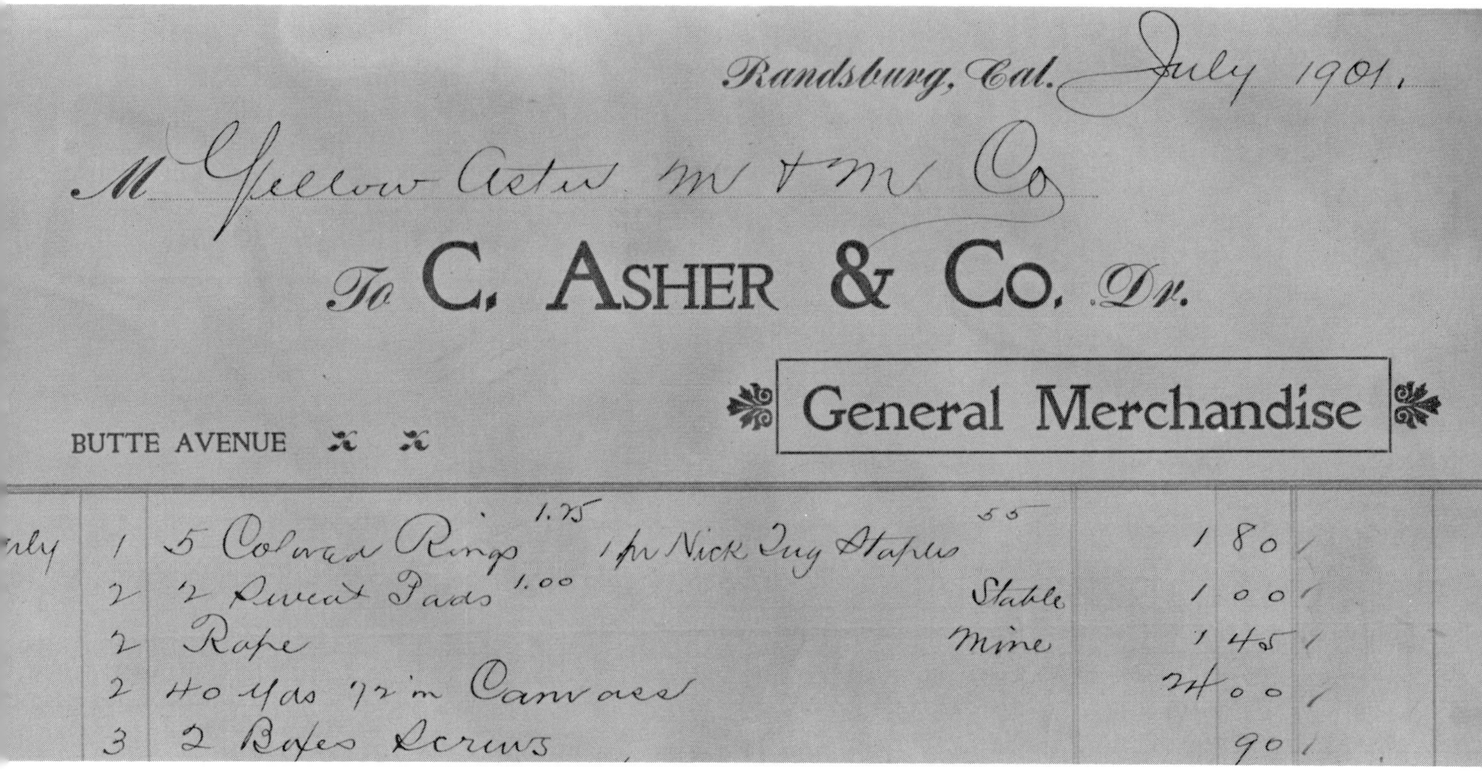

Randsburg, Cal. July 1901.

M Yellow Aster M & M Co

To C. Asher & Co. Dr.

General Merchandise

BUTTE AVENUE

July	1	5 Colored Rings 1.75 1 pr Nick Tug Staples .55		1 80
	2	2 Sweat Pads 1.00	Stable	1 00
	2	Rope	Mine	1 45
	2	40 Yds 72 in Canvass		24 00
	3	2 Boxes Screws		90

Yellow Aster bill for stable, mine, and mill items.

14

Financial Picture

An old account book and bills of the C. J. and E. E. Teagle Store in Johannesburg give prices during the 1910-15 period that show living costs for the period. Men worked a 12-hour day and ate lunch on their own time, for $2.50 to $3.50 per day. Even working six and seven days a week, a man did well to average $25 for the week.

The $25 did not go far, with butter 40¢ a pound, eggs 35¢ a dozen, bread 10¢, beef 35¢ to 70¢ a pound, a tin of oysters 50¢, a can of salmon 80¢, a tin of sardines 35¢, and coal oil for lamps 30¢ a gallon. The miner was averaging less than 30¢ an hour, with no paid vacation, no compensation if injured, and no unemployment benefits. It took him twice as long to earn a loaf of bread as it does today, but his liquor was cheaper, and the shot glasses were larger.

Working for himself, the miner often went long periods at a time with no gold to show for his work. The merchant trusted that a good ore body would show up and that he would be paid the mounting food bill. If the miner worked for a small operation, he was frequently expected to furnish his own candles (at $3.75 a box, more than a day's pay) or carbide for his miner's lamp. He even had to furnish his own fuze and dynamite in some instances at $1.30 for 200 feet of fuze and $7.50 for a box of powder. Payday was once a week, usually Saturday.

There was very little day-by-day food shopping except for those living right in the settlements; outlying camps, even Atolia for some time, were served by a once-a-week delivery wagon that picked up the next week's order at the time of delivery on the previous week's order. When tungsten was in demand in World War I, the

Teagles put a store in Atolia, the last of their chain of supply points that reached from Skidoo to Atolia.

Bill Bastion, who was hired to operate the new store, proved to be more interested in amassing a fortune on the side. He encouraged the Atolia kids to go out hunting "spuds," high-grade tungsten shaped like potatoes that lay on top of the ground in some places. Bastion paid them 25¢ a pound and then sold the spuds for as much as $4.25 a pound.

The Teagles were not aware of their employee's side business until the store was robbed one night and they went in search of Bastion. He was found running hoist on the night shift at one of the tungsten mines, and the store management job was finished. Shortly after that he was fired from the mine operation, too; it was discovered that he was letting tungsten nuggets spill down the side of the dump and paying boys 25¢ a pound to collect and bring them to him for pay the following day.

The Teagles were involved in many businesses other than merchandise stores and at one time were accused of owning or trying to own all of Johannesburg and to get a toe hold in Randsburg. They bought the *Miner*, which was published in Randsburg, but found it a time-consuming, unprofitable project. As Ed Teagle expressed it, "We were glad to unload it on Dr. Rose Burcham." Considering the business ability of Dr. Rose it is doubtful if she lost anything in the purchase.

The following year, 1911, just 15 years after the first milling of gold from the Yellow Aster, Dr. Rose was honored by being included in a list of the 20 most important women, selected by Elbert Hubbard. His selection was based on work known to have played an active part in the world's progress. That year she was honored in a Southern California souvenir book, *The Land of Heart's Desire,* published by the *Los Angeles Morning Herald*. The lengthy article about her stated:

> In spite of her glorious past, hers has been so much of striving and achieving and helping that it sometimes seems odd that she has gone so far, done so much, and still has time for so many things.

In the wordiness of the period, the article told of her grubstaking the men, the discovery of the Yellow Aster, and the years when

> Word has been quietly passed that a woman could hold her own along the frontier; her faith in mining; her long period of apprenticeship; her legal controversies; her final decisive victories; all these spirited chapters in her life have long made her a marked woman. The famous Yellow Aster Mines of Southern California, which have added $6,000,000 to the world's wealth, are inseparably associated with Dr. Rose L. Burcham's foresight, push, pluck, and enterprise. The story is one of the true romances of American life and Dr. Burcham's aggressive part reveals in a new way what a woman of clear thought and high purpose may do against seemingly insurmountable obstacles.
>
> Of $6,000,000 bullion shipped, over $1,000,000 has been paid to stock holders in dividends, $1,500,000 has gone into development, and the remainder expended in operating expenses. There is a huge body of 8,000,000 tons of ore blocked out averaging $3.00 a ton. It is above the mill level and is quarried or "glory-holed." The cost, while high, leaves a handsome profit.
>
> At the mines are eleven miles of railroad tracks, engines, locomotives, ore-cars, three water pumping stations, two wells, one 1,700 ft deep, the other 1,500 and impounding reservoirs of sufficient capacity to supply the needs of 100 stamps.

Dr. Burcham's social graces, "a polished lady of the drawing room" were now in a rightful setting:

> Splendid new villa in view of the mountains of southern California, is the realization of a long deferred dream of literary and social happiness. Her house, which is distinctive enough to be noteworthy, even in a land of magnificent homes and gardens, such as grace the perfect climate under skies of blue, has a splendid location on one of the gentle slopes overlooking the Sierra Madre Mountains.

The financial report of the Yellow Aster, made at the close of 1912, showed $120,000 paid in dividends for that year. The total since the start of operations amounted to $1,111,879. The company at this time held 40 patented claims, with ore bodies ranging in width from 4 to 30 feet, some of the smaller veins being the richest, with assays running $1,000 to $9,000 per ton. In most of the blue quartz ledges there was heavy impregnation of free gold in addition to the sulphides found in the big veins.

Most of the work was in the glory hole from which shafts and tunnels branched. A dynamo engine furnished electric lighting for the tunnels and underground development. Four locomotives hauled the ore to the 100- and 30-stamp mills and removed the overburden to the dump. Fuel oil was used in generating power for the various machines; two tank wagons with 14-horse teams transported the oil from storage tanks near the railroad in Johannesburg.

The mills were running continuously with three shifts, employing a total of 250 men. Water for the operation was pumped to the mills from Goler Wells in the valley to the north and Old Wells east of Randsburg. The Yellow Aster, in continuous operation for 17 years, held a time record equaled by few other mines in California or Nevada. Most unusual was the fact that it was still in the hands of the original discoverers, with John Singleton, President, C. A. Burcham, Vice President, and Dr. Rose Burcham, Secretary.

Many changes were in progress in the larger mine operations as the Nevada California Power Company, South Sierra Power Company, and Sierras Construction Company brought in electrical power late in 1912. Materials for a substation at Randsburg came as far as Johannesburg via Santa Fe; as the line moved into the valley to the north, supplies were shipped over the Southern Pacific line from Mojave and stockpiled at Garlock. Extensive work was planned for the Goler pumping station in addition to the towers and line needed to continue the high line on toward Bishop.

The erection of the steel towers that are still in use was no small task when working equipment was limited to manpower, horsepower, rope, blocks, and poles. A crew of 35 men dug holes and made cement piers for the towers. A 20-man crew assembled the towers on the ground, and then the massive metal structures were hoisted into the air with the aid of a gen-pole, using logs as rollers pulled by an eight-horse team.

From the time the towers were ready to be moved, the whole process was dangerous, but the men were skilled, and the few accidents that did occur were due to the rope breaking and whiplashing those in its path. Eight-horse teams pulled bunkhouses and cookshacks for the crew, following the advance group of workmen. When the towers were in place, another crew moved in to string wire, splicing at half-mile intervals. The line was run over rollers on the crossarms so that there was no bend in the soft aluminum wire. Splices were made by hand, a real skill considering the place, the equipment, and the fact that those splices have withstood high desert winds and quick changes of temperature for over 55 years.

The workmen found the Rand area going full blast, with all kinds of development under way and little unemployment. Men were working 9-hour shifts 6 days a week. Saturday nights were the highlight of the week with a dance, a show, or the bars with girls and gambling. W. J. Young, who drove a seven-seat, four-speed, 60-horsepower Pope Hartford for the power company officials roomed at the St. Charles Hotel in Johannesburg while construction was under way.

"We were at the St. Charles the night they had their grand opening," Young recalled. "That was a real shindig! After dinner the dining room was cleared for dancing to a one-record-at-a-time, wind-up phonograph. The evening was a lively one with 110-proof whiskey in quarts."

Young added that the installation of electricity at the Goler pumping station was a real aid to the Yellow Aster milling program as they had experienced some bad problems with gas engines. Guaranteed, continuous power was what they needed. Of the Rand acquaintances made during that period, Young said that "Charles Burcham was one you would never forget. He was a real likeable fellow, but seemed like a dude for this country. You'd see him in high calfskin boots and English riding breeches; then he would be in a white linen suit or other riding breeches. He changed clothes about three times a day."

As power line completion moved toward Randsburg at the rate of 3 miles per day, the local *Miner* noted:

> An area of 100 × 100 feet has been leveled east of Fiddlers Gulch, south of the main road between here and Johannesburg, where the sub-station will be located. Eighteen tons of material have been hauled to

the grounds and the erection of the sub-station will be under way in a few days. This distribution plant will be an open air structure, without roof or walls, covering a space approximately 80 X 75 feet. For air-break switches, towers about 30 feet in height will be erected, and the transformer in its waterproof casing placed upon a cement foundation.

Mining was in an upswing in 1912; the Atolia Mining Company had reopened after the previous year's shutdown and had seen the most prosperous year in its history. The price was on the rise, and approximately 100 men were employed in crushing, concentrating, and refining operations.

The King Solomon had for some time been worked only along an old ledge; then five groups of leasors went in and in three months had opened up several new and worthwhile discoveries, putting the mine back into the class with the better producing mines. Ed Shipsey, the owner, worked the main vein at the 250-foot level and discovered that above that level, good ore had been missed. From the surface down he was working $400-a-ton ore and the prospects continued to look good.

Teagle and Lamberson, leasing the east end of the King Solomon claims, had hit a good vein at the 85-foot level that was extending upward for at least 130 feet, with gold in quartz plainly visible. An early milling of 22 tons produced $45.60 to the ton with the next shipment expected to run much higher. Next to this leased area Drennan and Wickard were sinking a perpendicular shaft with difficulty because they had run into rock that slowed their progress. Cagle and Otteson worked along an old ledge believed to be a continuation of a rich stringer and had several sacks of ore ready for shipment.

Out in the Stringer district, long-worked mines were still producing, and the Magganetta that had averaged $160 a ton in gold for a number of years, dropped under $100 per ton in gold but came up with a bigger profit through tungsten concentrates that had been previously disregarded. The Santa Ana and Winnie Mines in the same area were steady producers, but showed nothing spectacular at this period.

In July of 1912 a huge body of ore was blocked out for development in the Good Hope, part of the famous Kenyon group. Development was done on the 200- and 250-foot levels with ore running $51.48 to $80.00 per ton and getting progressively better. The Good Hope, a mine in the southeastern part of the Kenyon claims, was an almost undeveloped mine until the Consolidated Mines Company purchased it in 1910. It was part of the property located by John, William, and George Kenyon three months after the discovery of the Yellow Aster. In the early days the owners frequently pounded out in a hand mortar as much as $50 a day; with this they bought tools and food and paid for the first milling, which brought $3,000. When asked why they didn't put up a mill so they could handle larger quantities, the answer was, "The gold is as safe in the earth as in the best bank and we'll take it out as we need it."

When the Consolidated acquired the Good Hope along with the Cinderella, Arum, Red Dirt, Bald Eagle, Coloratha, Agium, Lillian V., Lillian V. #2, Bing Hole, Poloillos, Little Wedge, Galveston, Twin Brothers, Four Hundred, J.I.C., Butte Wedge, and Coloratha Wedge in 1910, things began to happen. Development got under way and a compressor plant and stamp mill began pounding out the golden metal. By March 1912 the property had become an important producer with a 55-ton milling bringing in a $3,450 gold brick.

Less exciting than the figures quoted on gold milled but important to the mining communities was the county's announcement that library books would be available if space and a custodian were provided by the community. The county would pay transportation costs, but the people would have to provide for other expenditures. Daniel Gunderson, owner of the Butte Avenue drug store and book store, offered space in his book store. The first shipment had books entitled *Molly Make Believe, Following a Star, The Rosary, Master of the Wheatlands, Boss of the Wind River, Winning of Barbara Worth*, and other bestsellers of the year.

Butte Avenue was also the scene of an accident that caused the death of one of Cantil's early settlers. G. W. Munsey, homesteader on undeveloped desert land, had put in years of hard work to develop a ranch, running cattle and gradually putting land under cultivation. In February 1912 Munsey had completed business transactions in Randsburg and was starting for home when his horse spooked at a strange object blowing along the street and threw the rider, causing internal injuries from which the man never recovered.

That same year another Cantil resident faced death, but through his invention of a unique war device, was saved. Charlie Koehn, the hard-working little German who had developed a way station at Kane Springs when the first gold boom hit the area, was being harassed by claim jumpers. A group of strangers tried to run him off the claims he held on Koehn dry lake; when that didn't work they moved in armed men.

Koehn, with a natural bent toward inventions, came up with a mobile fort, made of two wheels and the shaft of a single-horse buggy. A sheet of boilerplate was secured above the axle, and a round log under it provided the protection Koehn needed. With a type of harness fastened to the shafts and over his shoulders, Koehn crawled and pushed the log and boilerplate fort ahead of him to within rifle range. Bullets whined as the gangsters' shots bounced off the metal plate or thudded into the log in increasing numbers as the strange vehicle moved slowly toward them. When close enough, Koehn opened up with an old pistol, and the gunmen left in a wild scramble to save their own hides.

The claims Koehn fought for proved worth the battle; he sold them for $30,000 to a gypsum company a few months later. Not until some time after the battle did Koehn become aware that a neighbor, John Cassou, had gathered other neighbors to join the "wild Dutchman" in his fight. They feared his fort was vulnerable from the back and stood ready to help if the battle got beyond Koehn's ability to handle.

Early that fall people for miles around Randsburg had an uneasy day as rumblings, grinding sounds, and a crash broke the desert silence. Men working at the Yellow Aster Mine were alerted and stood as spectators, watching, as 25,000 tons of ore and waste slid, rolled, and dropped with a roar into the glory hole. Since everything was being run through the mills, the cave-in would save the time and expense of blasting. Although telephone connections now reached from Inyo county to San Bernardino county, a distance of 250 miles, with Randsburg as a relay point for long-distance calls, people within a few miles of town had no way of knowing what had happened to cause the alarming sounds.

The towns of Randsburg and Johannesburg, modernized with electricity, telephone, and Western Union service, were less than 20 miles from Cave City in Last Chance Canyon where whole families lived in one-room dugouts.

A number of gold mines in Last Chance Canyon in the El Paso Mountains west of Randsburg were in operation. Many of the miners lived in shallow tunnels or earthen rooms where some previous miner had worked out a pocket of gold along an ancient riverbed. When Will Leach hit a good streak and needed help in shoring up the tunnel, his brother-in-law Wallace Lee Berry brought the family from Los Angeles and went to work in the mine. There were a series of dugouts near the road into the canyon, and Berry moved his family into one not occupied at the moment. Other families and single men were neighbors in the settlement that became known as Cave City. Seldom was a dugout left empty until the early 1940s, when gold was of little value and jobs elsewhere offered a better living.

The mining in Last Chance Canyon ran the complete range of gold-producing methods from hand-hammering of rock, dry washers and wet washers, to a 6-ton, wet-process concentrating machine. The processing plant was the most elaborate operation in the canyon area, and it attracted much attention. A gas engine brought water up out of a shallow, 20-foot well, and although there was always sufficient water for the processing, the drawback was the remainder of the plant, which seldom worked as planned. At peak periods a ton of ore was processed per day, but complete secrecy surrounded the value of the ore, and many doubted that the machine was anything more than a come-on for uninitiated investors.

In a nearby draw, the famed Hatfield, rainmaker supreme, was experimenting with one of his earlier rain projects. Hatfield built a dam in a small side canyon that drained into Last Chance Canyon. His plan was to create enough rain to wash down the side of Black Mountain, run through his canyon, and deposit gold against the earthen dam. On October 4, 1913, a cloudburst washed out Hatfield's dam and the 6-ton concentrating plant. No credit was given Hatfield, however, as cloudbursts and flash floods were not unusual at that time of year.

Though water was no problem in Fremont Valley or the El Paso range, the Rand Mountain side of the area continued to suffer from the lack of water, and even with all the wealth and modernization, no year passed without destruction by fires, many of which could have been controlled if there had been water. One of the worst in 1912 was reported in the *Miner*, October 26, following the total destruction of two dwellings. Mrs. Edith McGinn had lain down to rest leaving a kettle to heat on a coal-oil stove. She fell asleep and awakened in near suffocation, escaping the burning house but suffering arm and face burns. The south wind endangered the home of W. E. McLean. Several times the building caught fire, but wet blankets and a bucket brigade saved it. Not so lucky was the house of R. T. Clegg on the other side of the McGinn home, where

> Lack of sufficient pressure in the water company's pipes gave the flame a good start and the residence of R. T. Clegg on the west was soon blazing fiercely. There was sufficient time however to save all the household goods. Connection was secured with the high-pressure main of the Yellow Aster Company and saved other houses that were in danger.
>
> Mrs. McGinn lost considerable money in a trunk and the total loss of her home and possessions.

Under Supervisor Rinaldi's leadership, citizens of the community contributed $150, and the following night the Randsburg orchestra gave Mrs. McGinn and her two daughters a benefit ball at the Yellow Aster Club.

The *Miner* further commented:

> This fire again calls attention to the woeful lack of adequate means for fire protection, especially on the hill portion of town. A suitable reservoir could easily be constructed to furnish ample pressure. It is a matter that should interest property owners enough to take some definite action as a fire with the west wind blowing could obliterate the town.

The financial picture of the Rand District in the 1912 period would not be complete without a look at Johannesburg as it was described in a propaganda sheet hoping to entice more residents as Randsburg had done earlier.

> Nature has not dealt niggardly with this portion of the desert. No better site could have been selected for a town than that on which Johannesburg stands. It is comparatively level, the desert stretching away on its south side to Kramer, while the Tehachapi and Piute ranges of mountains border it on the north. On the east is a range of hills on which are the Pinmore, Val Verde, Alameda and other large mining properties and on the west side is a low range of hills which separates it from Randsburg, one mile distant.
>
> It is the terminus of the Santa Fe Railway, has water and electric lights, telephones and motor cars and all the facilities of a modern town. The business interests of Johannesburg are always brisk, and the trade with the camps throughout the northern part of the desert is greatly increasing.
>
> This town is connected to Ballarat, Skidoo, Wild Rose and other desert points with an auto service running three times a week from the office of Rinaldi and Clark. It is not saying too much by stating that the St. Charles Hotel, which was built to replace the old Hotel Johannesburg, is a structure that would not disgrace any city, built in the Mission style, it is replete with every modern convenience.
>
> The sound of the stamps of the mills at the Phoenix and Red Dog greet the ear of the stranger on his arrival, and is the first thing to acquaint him with the fact that he is in a gold camp.

The writeup also mentioned the climate as one of the greatest natural sanitariums on earth; great wealth awaited the investor; and for the man without much money there were mines to be leased so that he could be his own master, working for no one but himself. It ended with the remark:

> Capitalists are beginning to realize that no matter how good the investments are in other branches of mining, there is only one metal whose standard remains unaffected by war or commercial depression—Gold!

R. S. Hamilton on right, opening a new pocket for the Yellow Aster Mine, 1909. *(Courtesy Pearl Langdon Norell)*

Randsburg business district, 1940s.

15

Some Fun

In the area now occupied by the general store and post office in Randsburg, a dentist, Dr. Baxter from Lone Pine, had a small office for the days he served the Rand community. As one of the boys of that period recalled, "Everyone that had had a toothache since the last visit of the dentist was hustled down to his office. By the time the doctor arrived in town the tooth quit hurting, but you had to go see him anyway."

All in fun on a hot desert day. *(Author's collection)*

Charles Siebenthal explained that "You were quite a hero if you went in to the dentist and the kids waiting around outside didn't hear a sound out of you. It was great being respected by all the kids after it was over, but at the time you climbed up in the chair and he put a paper bag over your head, you wished you were dead. The gas he shot under the bag soon dulled your desire to escape and then the whole thing seemed pretty funny and you didn't care what happened."

Next door to the dentist office was Grandpa Spencer's Ice Cream Parlor with the curled wire table and chairs, where ice cream was piled high with nuts and a cherry on top; colorful candy sticks showed through the glass of tall jars. Next door was Gunderson's drug store with rows of medicines, bottled and boxed.

For a short period a Dr. Garvin had an office along the same street. "The women went wild over him," one early resident remarked, adding, "He was handsome, looked like General Pershing, and drove a classy rig with a beautiful white horse. Though there were always 10 men for every woman in a mining community, that doctor had more than that number of women interested in him!"

A little farther up the street was Rott's Bakery, sold to Mrs. S. Z. Morgan in 1918, and then to Spaldings, who converted the building into a saloon now operated as The Joint.

Up the hill where the American Legion Hall and the high school were built in later years was Johnson's stable. Johnson ran a two-horse stage to meet the daily train that arrived in the evening. The stage, an eight-passenger affair, seated four individuals on each side facing the middle. The old road past the Big Butte Mine and over the hill to Johannesburg was the main road between the two communities. Usually the trip to the railroad station was uneventful, but one cold January afternoon in 1914 the team suddenly reared up, made a quick turnabout, and headed back for the stable. The sudden turn of the stage threw Johnson out into the sand and sage. As the horses raced into Randsburg, kids coming from school sent up the cry, "Runaway team, runaway team," and started following the jolting, rocking stage. Mike Hellen caught the animals before they hooked the rig on the corner of a building, turned them around, and drove out of town in search of Johnson, who was sitting, stunned, right where he had been thrown. The train was met on time, and incoming passengers were not aware that they were riding behind a team that was still a bit spooked.

About this same time the Rand district found some amusement at a shooting episode. It was common knowledge that the J. P. Cuddebacks had some lively arguments, as many of them took place right behind the bar of their saloon, regardless of who or how many patrons were present. Verbal insults frequently made an amusing evening for the customers, and when it was especially good, other bars lost trade and the Cuddebacks' place was full.

One night Mrs. Cuddeback backed up her arguments with a shotgun flourished under her husband's nose. In that expressive and explosive atmosphere the gun went off and hit Timothy Danahar. The wounds were painful but not serious. Mrs. Cuddeback was placed under $5,000 bond for attempting to murder her husband. Cuddeback immediately put up the $5,000 and took his wife home, saying that it was just an accident, that she really wouldn't shoot anyone.

There was fun outside the bars too. Mabel Edsel, who came to Johannesburg as a young girl and clerked in Teagle's store, still enjoys the memory of dances held in the Union Hall and the Lodge halls of Randsburg. There was a dance nearly every Saturday night and frequently more than one, so that they competed for the girls. At times Union membership dictated where the men danced; if the Union Hall had a dance, its membership was confined to that dance. "Since the two halls were back to back, it wasn't unusual for Union boys to slip over to the other dance, especially if the girls at the Lodge dance struck their fancy. No girl ever had to sit out a dance, as the men far outnumbered the women."

As popular in the daytime as the dances at night were the baseball games. Each town was loyal to its team, and the current issue of the *Miner* would try to be impartial as it gave a rerun of the previous Sunday's game. Typical was the front page, center column of the August 16, 1913,

Randsburg-Johannesburg baseball team, 1912. Gene Ward, pitcher; Herb Krause, catcher; Bill Dwyer, umpire. *(Courtesy Mabel Edsel)*

The Yellow Asters took the Randsburg-Jobergers into camp with a 7 to 6 victory on the Joberg diamond last Sunday in the first of a series of games.

Taylor entered the box for the Yellow Asters and things looked easy for Teagle's binglers in the first inning when Kinsey got a safe one and Krause walked. Rinaldi knocked an easy one to Taylor, but by the time Taylor recovered the ball all were safe. Sewell, the next man up, also took a walk and Kinsey was forced home. The last time Ward swung his bat in the first frame, Taylor was facing a ticklish situation. Men were on first, second and third and any kind of a single would have given R.J.'s a good lead, but the Cincinnati bingle was not forthcoming and Ward retired to the bench, Fahey following his example.

Ward was none too steady in the box which was a rare thing for Gene and the Yellow Asters made merry with him. The cause of Ward's uneasiness was the man behind the bat, Sewell, the new discovery from the Stringer district, who had better been left in obscurity.

The Yellow Asters were there themselves in the first inning and might have piled up a bigger lead save for the inability of their players to wallop the ball.

Kohlmeyer, whose forte is getting in the way of the ball, was hit by one of Gene's slow ones, but was not crippled for life, for he got to second and came home on Well's hit to center field who in his turn was brought home by McDougal.

When the Trona Railroad reached the Southern Pacific line at Searles Station in 1914, they sponsored an excursion to inaugurate the line. The celebration started at Randsburg with cheers filling the air as 50 joy riders started out in a large American Trona Corporation truck. The dirt road was badly rutted, the ride was rough, and the truck was so crowded that some of the fellows at the back were spilled out on the trail, adding to the fun of the trip.

At the Searles Station, the group boarded the special train, first official passengers on the new line. The 31-mile trip in the open-windowed coaches was a delightful, picturesque tour. Dinner was served at the company hotel, and then the Randsburg boys proceeded to win the afternoon ball game. After supper the group took the train back to Searles and the waiting truck. It was late when the worn-out inaugural participants finally reached Randsburg.

Box socials were always enjoyed, the girls secretly decorating boxes filled with a delicious supper and then sitting tense and waiting as the boxes went up for auction. Sometimes small signals were passed on to their favored suitor so that no undesirable man would get their box, since they would be obliged to eat supper with the buyer. The Methodist church came up with a unique box supper auction that left no chance to tip off the bidder. The decorated boxes were all stacked on a table; the bidding was not on a decorated box but on a nose showing through a hole in a sheet. There was no way of knowing which girl would be next in line behind the sheet, and the nose gave no clues.

Holidays were special for almost everyone, even if wages did limit the celebrating. Christmas may have been skimpy for adults sometimes, but the Rand area tried to make that time of year something to long remember and cherish. Clementine Lovett Elton spoke of her fondest childhood memory as Christmas in Randsburg. "It was an exciting time of making things, decorating, and practicing for a part in The Program. It seemed as if Christmas Eve would never arrive."

By December 24 the excitement had reached such a pitch that the waiting was torture. Adults seemed to want more time so they could get more done, while time moved far too slowly for the children. "At long last everyone was dressed, ready to go to The Program, and then I did wish the time for The Program was just a little farther off. Maybe it was a touch of stage fright."

The Program was held in the Methodist church. Early arrivals got the best seats; and the kids had time to move about, admire the tree, and share exciting little whispered nothings and nervous giggles. Mrs. Elton mused, "There never was anything as wonderful as those Christmas Eves."

The evening gave each child a chance to display his talents in song or reciting poems and readings. Home-made costumes would appear, as they had the year before and the year before that. Little Wise Men played out the scene of travelers to the manger, tripping over trailing robes or mismanaged shepherds' crooks. Being a performer gave a special glow to the whole evening; the recitations may have been dull for the adults but to the child it was all exciting and important. "A high-grade performance was expected in spite of forgotten words, a verse skipped, or the younger brother or sister running up to hold your hand and share the glory as you worked your way through snow, sleigh, and Santa poems."

There was always a delay following The Program. Children grew restless with waiting and moved about until there was the sound of sleigh bells on the desert air; then everyone would rush back to their seats. "Now it was really the night before Christmas, with not a child stirring or a sound made as a big fat Santa Claus crawled through a side window and men helped him pull his bag of gifts inside. Santa's booming "Merry Christmas" broke the spell and was answered with a similar greeting that sounded much like a cheer.

"Christmas Eve was the only time we saw Santa and the only time we received gifts except at home." Every child at The Program knew that he would receive a gift in addition to apples, oranges, and hard candy. The town saw that every child was there that night; otherwise some would have had nothing, not even an apple for Christmas. It was not until they grew older that the children realized that it had been the townspeople who had seen that Santa's bag was well loaded and that the "girls of the line" had been the largest donors toward the gifts. "Though the girls were looked down upon by many of the citizens of the mining communities, there would have been any number of children and whole families that would have had a mighty skimpy holiday if those same ladies had not provided the needed cash to buy a doll and fill a food basket."

Not all the fun was in the towns. Even the sparsely populated areas had their brand of amusement, like the spring day when a carload of tourists brought their steaming vehicle to a stop near a prospector sampling the sandy wash near the rutted road in Red Rock Canyon. As the driver called out a greeting and asked where the road went, Bill Stevens, who lived in Cave City in Last Chance Canyon, looked them over from the ladies with big hats covered with dust veils to the gloved, goggled, and coated men. In calm dignity he replied, "Keep right on out of the wash; turn right and go to New York; left and you go to Alaska. You can go anywhere you want to from here." Then he turned his back on them and went on with his search for gold.

Girls returning from Fourth of July celebration pause in front of the Burchams' first home in Randsburg.
(Author's collection)

Building Los Angeles Aqueduct through Jawbone Canyon, 1908-09. *(Courtesy Stover Lowe)*

16

Changes

Few people in the Rand area could believe that anyone would want to settle out along Cuddeback dry lake. But early in 1913 Golden Valley, locally referred to as the Cuddeback district, was opened to homesteaders, and families began arriving who were not the least interested in mining. They were going to carve farms and ranches out of the desert lake bed. Claims were filed, and a full-fledged colonization project was under way.

A March 1, 1913, report reads

> An extensive acreage has been cleared of greasewood and brush, many wells have been put down, ranging in depth from 60 to 125 ft. Houses and barns are being built, fruit and eucalyptus trees set out, and acreage has been planted to barley.

Locally the project was viewed with skepticism. Sunday drivers took families out that way to welcome the new settlers and get a look at what really was going on. They were surprised at the number of families already settled and the immense amount of work that had been accomplished in about two months. Some of those working their homesteads and the extent of their development as listed in the *Miner* were these:

> John A. Johnson, 160 A.; 40 A cleared, 10 A sowed to barley. Small acreage to fruit trees, 5-inch well, 70 ft. depth.
>
> John A. Trobeck, 160 A; 40 A cleared with barley and fruit trees set, 5 inch well, 85 ft. depth.
>
> Mrs. Ostrander, 160 A, land is being cleared.
>
> Nels Johnson, 160 A; house, 40 A cleared, 20 acres plowed.
>
> Leopold Arnt, 160 A, 3 room house, barn, 40 A cleared and seeded to barley, 12 inch well, 120 ft. deep.
>
> D. A. Morse, 320 A, small barn, 40 A. cleared and plowed, small acreage set with eucalyptus trees.

G. H. Atchley, 320 A.; 65 Acres cleared.

. . .

H. Rowantree, 160 A, house, garden, 20 A cleared, 6 acres seeded to barley, small acreage in potatoes. 5 inch well, 68 ft deep.

. . .

Albert J. Burke, 160 acres, house, small acreage set with peach, apple and cherry trees; 5 inch well, 105 foot deep, 35 feet of water.

Anna Kreidt, 160 A, house, land being cleared.

When the April winds came, the cleared, plowed land moved east; young barley shoots were twisted off at the surface of the ground or blown out of the dry, loosened earth. Some people left, never to return; others looked for jobs to support their families and worked at clearing and well digging in their spare time, realizing that an irrigation system had to come before any type of crop could survive. Some hand-watered trees clung to life, but generally for those planted before a well was developed, the effort was a total loss. New homesteaders arrived with high hopes but little knowledge of desert farming. The few who stayed on, meeting government requirements for proving up on their homesteads, found employment in the mines or outside support.

In July 1913 Kern County held its first road bond election, and $53,850 was allotted to the Rand area to improve and build roads. Cars had been seeping into the area over dusty, sandy, rutted paths. A Model A had replaced the stage to Ballarat; trucks were making short hauls. Better roads spelled the end of the long-team freighting outfits; only old experienced drivers kept the more isolated places supplied; no new, young drivers were being trained.

Change, other than modernization, hit the great Yellow Aster Mine Company, too. Charles A. Burcham, best liked of the three men who discovered the gold on Rand Mountain, passed away in August 1913. The local *Randsburg Miner* ran a lengthy salute to the man, describing his childhood, business ventures, and mining activity, and closed the account with this statement:

He was member of the Benevolent Protective Order of Elks, Odd Fellows, the Jonathan Club of Los Angeles, the California Club and had a host of friends who admired him for his ingenuous disposition, genial good nature, and sterling character. Besides having been one of the main stockholders of the Yellow Aster Mining and Milling Co., he had other extensive mining interests in California and Arizona. He was president of the Rand Mercantile Company, the Rand Water Co., and the Randsburg Telephone Exchange.

Nine months later, in May 1914, John Singleton died, and the *Randsburg Miner* carried a short paragraph boxed in black. It said that he died at the age of 68 and left a wife and brother. The estate amounted to $215,648. Included in this were 18,241 Yellow Aster shares valued at $200,000; the Windy Mine, $10,000; and the Kathy Ross in Nevada, $2,500. His wife, Stella, living in Europe was the principal beneficiary.

Dr. Rose Burcham was now the only remaining member of the original ownership of the Yellow Aster properties. She watched inheritance litigation eat away at the very foundations of the organization and skim off the profits from the operation.

Six months after Singleton's death the *Randsburg Miner* printed an even shorter notice, just one line saying that Barney Osdick, aged 47, had died from an explosion. The community had reacted more deeply to Osdick's death than to that of the President of the Yellow Aster Mining and Milling Company, and there was no need for an explanation in the paper, for the story of the tragedy had been told and retold.

Barney, a long-time, well-known miner of the area was working at the jinxed Phoenix mine on the edge of Johannesburg. Witnesses were unable to explain how Barney's actions had set off the explosion. Donell and Elingsworth, Randsburg merchants who had leased the mine, were making a changeover in processing. While the new plant was being built, Barney Osdick was hired to coat all iron areas against chemical action by painting them with P & B. The covering material was a mixture of asphalt and Hi-Life (carbon disulphide), a highly explosive combination. Barney, wise in the use of P & B or any other explosive material used around mining, was having difficulty with the thick, hard-to-spread stuff on a cold morning. Witnesses were hard put to explain why Barney of all people would build a fire under the barrel of P & B and then stir the fire into a blaze with a stick. As the stick blazed up, the barrel exploded, killing Barney—another black mark for the Phoenix.

One of the most desirable changes for the Rand area seemed to be taking shape now. The communities on Rand Mountain had suffered great losses and discomforts from the lack of water, but at last the problem was to be solved. In March 1914 the *Randsburg Miner* headline announced: "Golden Valley and Rand Communities to be turned into a Veritable Garden." No more water bills that based the billing on

Family rates $3.50
Bachelors: one $2.00, two $3.50, three $4.50
Roomer: $1.00 extra
Bath tub and shower bath: each $2.00
Desert cooler: $1.00
Family doing laundry work: $7.00
Saloons: $10.00
Horses: $3.00 each
Cows: $3.00 each
Garden hose prohibited, no gardens allowed,
Outside faucet prohibited.
Parties giving water away or selling, will not be furnished water.

For the solution of the water problem, A. P. Appling and Thomas M. McCrary, Surveyors, and Robert Conley, Civil Engineer, had prospected the valleys surrounding the Rand area and found abundant water at depths of 35 to 100 feet in the sandy loam of the valley between Randsburg and Garlock. Sixteen sections had been secured and contracts let for 15 wells. The old ore road to the Garlock mills ran through the property, the sandy road over which thousands of barrels of water had been hauled from the Garlock wells to the mining community on the mountain. Notices were posted, encouraging homesteaders to take up what land was still open, promising them astonishing results as soon as the water was applied to their land.

Results were not that magic. One by one the wells were capped and forgotten. Lack of community interest and financing killed the project, and no further attempt was made to get private financing or community backing. Families continued to haul their water as the supply coming from the springs and wells east of Red Mountain had never met the demand. Stewart Fraser, who managed the water system for many years, said that the large hand-dug wells with shafts down more than 400 feet at Squaw Springs and Mountain Wells, 10 miles from the mining communities, furnished enough water for drinking but not enough for other purposes. Mines like the Sunshine ordered water by the tank car, brought in on the Santa Fe, and piped miles up to their mine and mill operations.

The water wells, shored up like a mine shaft, were large enough for gas engines to operate below ground level with connecting booster points in order to get the water to the surface. At one time a short developed in the wiring and set fire to the timbers; there was no way to keep pumping water out of the well and there was not an adequate reserve in the storage tank for fighting the fire. Santa Fe hauled ten tank cars a day from Barstow for fighting that fire that took two weeks to smother.

Two children who continued to haul their family's water supply from Garlock to Randsburg were John and Lizzie Robinson (Short). Their uncle, John Beatty, still had a slaughterhouse at Garlock, and they hauled beef to his meat market in Randsburg and water for their own use. They filled their barrels and watered their horses at a well near Granny Slocum's. "We drew water by hand via a series of buckets attached to a rope and filled buckets at the side of the well for the horses." John added, "Kid-like, I started to run the buckets as fast as I could and suddenly the whole outfit that made one revolving bucket system and guide for the rope, crashed to the bottom of the well. The old lady never did like kids much, and when those pails hit bottom, we were scared. Lizzie and I jumped in the wagon and started the team for home at a gallop."

Sarah Slocum or Granny Slocum as she became known to most, arrived in Garlock in 1914. Land was cheap; the town was dead; owners of lots were happy to sell for any amount. Within a short time she had accumulated a good portion of the town site and was leasing out buildings to the lone miners who remained, still trying to eke a living from the canyons. Some of Granny's descendants said that she came from New York to San Francisco, accumulated $14,000, and then moved on to buy up cheap desert land.

The southwest end of the Rand area was increasingly tending toward agriculture and less effort went into mining. Homesteaders around the present Cantil were at an all time high in 1915, through the war years. The nation was encouraging food production, and good tillable homestead land had already been developed; there remained the unconquered arid regions like the Mojave Desert. The sand was ideal for some crops, if it could be made to lie still when the spring winds blew. Water sources could be developed and the sage cleared without too much difficulty. Land made available for homesteading started at the eastern edge of California and took in the Lucerne, Antelope, Fremont, and Golden Valleys. The requirements were minimum: a well, a dwelling, a sown crop on cleared land, and "in no time at all the land is yours." It sounded easy and wonderful, but few stayed to prove up and become owners.

Around Cantil the more courageous clung to their dreams, fought the elements and helped each other. Among the hardy few that remained were the Cassous, Pappus, Munseys, Van Winkles, Burdicks, Blalocks, Giddings, and Todhunters. By cooperative effort they built homes and a school. The Red Rock Improvement Club, with membership open to everyone in the area, had Saturday night get-togethers and community planning sessions. The Club was credited with building the school and contributing the most to the social life of the struggling community.

Mrs. Levi Gidding, member of the first school board, recalled: "It was a hard, discouraging life and the Saturday night meeting of the Red Rock Canyon Improvement Club or an all-day house raising followed by the Club's social evening was the only pleasant spot in many weeks of otherwise bleak existence." The social evening started with a quick, to-the-point business meeting, and then the school children would "recite pieces." Group singing was a favorite kind of entertainment, and occasionally games occupied the remainder of the evening. But most often they would talk Todhunter into spinning a yarn that captivated all ages.

Building the school was a monumental undertaking for a community that was struggling to hang onto the land and still keep food on the table. Fred Hart donated the land and the material while other men donated their labor. Willard Burdick, Will Munsey, and Mrs. Gidding made up the first school board; their task was greater than their titles indicated. The school board sponsored every possible means of raising money for equipment, hired the teacher, and finally canvassed an extensive area to encourage parents to bring their children whatever distance was necessary so that they could take advantage of the education offered. It took two years, 1914 and 1915, to get the project on its feet, serving all eight grades for students that varied greatly in age within each grade.

Cantil did not exist when the first homesteaders arrived, but there was a small adobe building used for groceries and mail to the west of the present Cantil; Kane Springs was still a supply point to the east. Jack Harrison operated a small store along the old Neuralia road for a short time before he opened his famed saloon in Johannesburg. The Giddings homesteaded just south of the abandoned adobe store, putting all tillable land under cultivation. There was so little return for their efforts that Gidding went about the country doing mine assessment work for absent owners in order to have grocery money. Other men in the area were doing the same or working on the road and building for those who could afford to pay for such work. A few were able to pay to have a well dug, but many relied on the help of others; some, like Pete Cassou and his wife, pulled, by hand, a small cart with cans to fill with water at a nearby spring.

Tall, slender C. C. Todhunter, who took up his 160-acre homestead in 1912, not only had a way with storytelling and quick wit, but could get plants to hang on and produce regardless of climatic conditions. A windmill pumped water onto the sand, and his farming efforts progressed. Watermelons from his 5-acre plot found a ready market in Teagle's store in Johannesburg. As soon as Todhunter's wagon backed up to the dock at the store, word got around and the townspeople came to buy before the melons were unloaded. "They were one of the prized treats on the Fourth of July."

A. J. and Clarence Crookshank of the First National Bank, Santa Monica, had more grandiose ideas for the use of Cantil's desert. They invested their wealth from the bank and from a sugar-beet industry in Orange County in 5,000 desert acres with plans of planting them to sugar beets. The oil craze hit, the sugar beets were forgotten, and the fortune went into six dry holes. The whole Crookshank project was abandoned in 1927. Today the Thunderbird Ranch raises Peruvian Paso horses where the sugar beet empire was to have been built.

Though many of the Cantil settlers had little use for mining as a means of making a living, regarding it as too great a gamble, they did admit that sometimes it paid far better than did their farming efforts. Early in 1915 gold nuggets were making news throughout the Rand area. Dave Bowman discovered one that had all California talking when it sold for $1,979. His big find was made in Red Rock Canyon about the same time Al Wiser and Charlie Koehn brought an $800 nugget out of their mine near Red Rock, and Archie Martin picked up a $500 one near the mouth of his mine located between Red Rock Canyon and the Last Chance Canyon road. The news of nuggets set off a new surge of prospecting, and men who had let their claims go unworked for years started to dig again.

At the time Dave Bowman's $1,979 nugget was being sold, the old Sunshine Mine near Randsburg was again making headlines. Twenty-two tons of ore had been mucked out following a blasting session, and free gold glittered in every piece of rock. The milling showed, per ton, $54.50 in free gold and $900 in concentrates. This mine was being closely watched by the mining world as a possible new "Mother Lode" and another bonanza for the Rand.

The Sunshine had been located in 1896 by B. M. Atkinson, father of the 1915 owner, Tom Atkinson. Located about a mile out of Randsburg in the Stringer district, the first shaft went down 200 feet in 1898; at 30 feet a 6-inch vein was struck that produced several hundreds of dollars to the ton until the vein pinched out. When the Bully Boy and the Merced Mines were added to the Sunshine, the income was great enough to enable the owners to install hoist machinery and a compressor so that work could be done on a larger scale. Most of the $500,000 taken out of the Sunshine went back into development, working nine or more men in the mine, a three-stamp mill, and a concentrator at the time the $1,000 per ton ore was hit.

Though no new boom hit the Rand area, the mining towns were busy, prosperous, and optimistic until the mining strike of 1916. Business fell off, there was an uneasy atmosphere, and employment was scarce. Money was the tightest it had been for years, but two enterprising brothers thought it a good time to go into business. They took over the White House Saloon, liveliest place in town that provided food, liquor, entertainment, and girls. Their ad in the local paper added class:

> The White House Bar
> E. J. and Mose Elder Prop.
> Wines, Spirits and Liquors of Fine Quality
> Hot Tamales and Sandwiches Day and Night
> Best Selection of Cigars in Town

The food undoubtedly came up in the big dumbwaiter from the White House Cafe in the basement, managed by Tom Garrity. The restaurant below street level was confined to one side of the excavated area; a massive refrigerator-ice house was built on the other side. Narrow tracks carrying small wooden ice cars took the cold load from the ice room to a back door that came out on a sublevel opposite the row of floozy barns.

The Elder brothers were ambitious business men whose wholesale ice business served a wide desert area. Next to the underground restaurant was a wine cellar of thick adobe walls under the dance floor; here they stocked the wines served up at the bar and sold through their wholesale wine distributorship. Barrels of wine were stored, bottled, and sold to local merchants. Rather than make trips about the country drumming up business, Emmit Elder set up a "first-class" mail and passenger service to Skidoo and made regular calls on his customers in Trona, Ballarat, and Skidoo at the same time. His first class passenger service was in a Model T with gas tank mounted above the dashboard so that he could get up the slopes. A tire pump attachment helped push the gas through so he could get over the steep Slate Range that had to be conquered on each trip in and out of Panamint Valley.

According to Dr. Rose Burcham's niece, Maude Bushman Noe, who worked in the Rand Mercantile at the time, the man who came into the store dressed the neatest, had the whitest teeth, and was the only one always to pay cash was Mose Elder, the less aggressive, smaller spender of the Elder brothers.

As the nation moved into World War I, many of the young men were called into service, but the seasoned miner's skills were needed more in the tungsten fields of Atolia. It was a boom time, with the ore in demand for the war effort; approximately 4,000 people worked, lived, or circulated in and out of Atolia in 1917-18. Money flowed freely as did the 4-year-old bourbon from a charred oak barrel at 10¢ a shot, the third one on the house. High-grading was an accepted fact and only when it became too open or too profitable was any claim made against an individual. However, there was a shuffle in management when one mucker banked $80,000 in one four-month period.

For over 20 years the local newspaper had been the *Randsburg Miner*, carrying news of the Rand area communities and mines as well as mining items of California and Nevada. In June 1917 it appeared as the *Golden State Miner*, under the new ownership of Watson and Bowsher. The masthead was changed to read at the left of the name, "For Big Men Who Can Command Big Capital" and on the other side: "Home of The Yellow Aster Mine that produced a ton of gold every year for 20 years."

Registration Day in Randsburg showed Precinct #1 with 64 registrations (28 Americans, 36 aliens, 15 dependents, 12 defectives with disabilities, 12 with no dependents). Precinct #2 had 28 (24 Americans, 4 aliens, 2 had dependents, 2 defectives, 2 no dependents).

At that period the advertisements in the *Miner* read like a directory:

White House Bar, E. J. Elder
Palace Bar, Butte Ave.
Old Crow Bar
  Makes the old grow younger.
Palace Bar, Royal and Phillips
  Pocket Billiards and Barber Shop
Houser Hotel, D. E. Thompson, Prop.
Rand Hotel, S. A. Olteson, Prop.
Belts Hotel, Mrs. Bernice Belts, Prop.
  Regular boarders $35 per month
Rinaldi's Garage (Hill Bros. old stand)
St. Charles Hotel, Teagle Bros. owners, mgrs.
Bill the Junkman, J. W. Rosswurn, Atolia
Ruffhead's Barbershop
  Shave, haircut, shampoo, bath.
Peter Christensen of Maginnis Hotel
  Gone back to old price,
  Meal 35¢
  Steady boarders $1.00 per day.

Just prior to the opening of World War I, the Cole brothers, Darrell and Rolland, returned to Goler where they had prospected three years earlier. This time they brought in an engineer who wanted to try hydraulic mining at the long-worked Summit Diggings. Four-inch pipe, shipped in from Germany, was laid from a well they dug in Goler, to Summit over six miles away. The project cost about $40,000 and production returns were not paying for the investment even before the war put an end to the operation.

South of Goler the idle gold mills of Garlock were being dismantled and sold for scrap metal by Granny Slocum. The price for the scrap was good, and there was no one around to question the sale. J. D. Voss, a stocky German mining engineer, had moved into Garlock, but his interests were directed toward promoting a new settlement to the northeast in Goler. The town he was building was to be known as Vossville, a center of activity for his sale of mines scattered throughout the El Paso range.

As suddenly as the smallpox had hit in 1901, the flu epidemic of World War I took its toll in the Rand area. Dr. W. L. Denton of the Yellow Aster Hospital tried to be everywhere and care for everyone but it was an impossible task. His speeding Pierce Arrow was a familiar sight; road conditions made no difference in the rate of travel. There were those who swore they wouldn't ride with him, if their life depended on it.

Pete Osdick, resident since 1896, owner of a mine and mill between Johannesburg and Atolia, performed outstanding service during the height of the epidemic. He ministered to the ill and cared for the dead and once wired the governor to ask for additional medical aid, or even for some one authorized just to sign death certificates. The record shows that he officiated at 19 burials in 10 days. Unofficially it was understood that those well enough to attend the services were to be fortified by bottled spirits and by a face mask with three holes; two holes were to see through and the other was for a cigar to keep up a smokescreen to kill the germs.

As suddenly as the influenza hit, it stopped. Men began returning from military service; the first ones home found employment; later ones found less opportunity. Demand for tungsten dropped, and many of the jobless settled for little more than board and clothes as they worked to develop other men's claims or struck out to prospect for gold on their own. One of the returning soldiers went to work for Charles Knight, Justice of the Peace in Johannesburg, on his Colorado Camp gold mine facing Black Mountain. "Frenchie," C. C. French, and Colorado Camp became synonymous as he later became owner of the claims and camp buildings. The deposit was none too good, but every shovel of dirt showed some color; it seemed more useful for interesting investors than for an actual mining operation. Frenchie developed promotional material for selling claims, but couldn't meet the requirements of the law and was never able to make a big sale or make a lot of money the easy way. He carried a fair-sized nugget around with him and talked numerous individuals into grubstakes from $25 and up per month. With money in his pocket, he and his short-haired white dog, Bow Wow, would get into an old truck and head for Mojave to pick up supplies. Since he couldn't resist a game of Coon-Can, as little as possible was spent on gas and food. As long as the money lasted, Frenchie played; then, flat broke except for the food in his truck, he would start back to Colorado Camp.

On the way out from Mojave, Frenchie would usually pick up a hitch-hiker and the conversation would get around to gold and his need for a helper. He would feed and go fifty-fifty on the gold dug out; claiming poor health, Frenchie would stay topside and run the drywasher while the fellow did the underground digging. Each day there were some colors to share, but "We must be in a poor streak," was the word. This process went on as long as the hitch-hiker was willing to work for food, a roof, and a few specks of gold a day, or came to realize that if there was any sizable gold, it was being lifted from the drywasher while he was underground.

In slim periods between grubstake angels and diggers picked up off the highway, Frenchie pecked out page after page of gripes on an ancient typewriter. His tirades covered everything from weather to government, but mostly complained about taxes of which he paid little, having no record of income and living on a nontaxed mining claim. Frenchie's writing went to the *Saturday Evening Post* and in due time was returned; each time he failed to break into print, an acid letter went to the editors accusing them of either being a pawn of the government or cowards.

For years Frenchie survived by using the gullible, playing on the sympathetic, and overglamorizing his gold claims, but he was not the only one. Many made a living by different techniques of "salting" their diggings, their glowing descriptions of gold returns, or sharing a claim for money to be used for further development. The game was the same; only the methods varied.

Kelly Mine main shaft and mill buildings.

17

The Big Silver

In 1918 the Rand District started on a third mining boom. It was gold in 1895, tungsten in 1905, and now one of California's largest silver discoveries.

One day, Hamp Williams and Jack Nosser, both well-known miners, were walking the area south of Randsburg collecting clay samples for possible mining and sale to a clay products company. On the return trip, heavily loaded with their collection, the two men paused to rest, and an unusual blue-gray outcrop drew Williams' attention. Breaking off a chunk of the odd-appearing formation, Hamp Williams brought in a new strike of high-grade silver.

For years men had walked the very same area in search of gold and found little or nothing; the Osdick brothers barely made a living out of the gold they mined a short distance away. No one had looked for silver. They had tramped over a fortune because they could think of nothing but gold.

John Kelly, Kern County Sheriff, who was associated with Nosser and Williams in the proposed clay project, was immediately informed of the silver prospects. Financing would be needed if mining operations were to be undertaken; while Nosser and Williams staked 22 claims, Kelly was busy in Bakersfield getting his friends interested in buying into the new Kelly Mine. The shareholders' money, plus what Kelly and Edith Coons, Kern County Assessor, could put into the project, amounted to $4,000; this financed a shipment of ore that was all surface rock.

That first carload of ore sent to the smelter produced $2,300 net, and the next one $4,700, with no underground work. The first shaft netted over $200,000 to the 90-foot level; in the next four years 60 major shafts were dug for deeper, richer ore. The Silver Glance, Silver King, Santa Fe, Big Four, and other, smaller operations were also striking good silver, which indicated that the ore body extended over a considerable area.

The discoverer of what came to be known as the Rand's Big Silver, Wade Hampton "Hamp" Williams, was the son of the Hamp Williams that opened up gold in Havilah country in 1861. Both men, carrying the same name, made great discoveries just 58 years apart. Though young Hamp was raised on a cattle ranch in Kelso Valley, prospecting ran in his veins. At the age of seven he found a gold deposit that he sold for $100 and a horse; that deposit produced over $100,000 in gold through the years. In 1900 the drought depleted the family income and Hamp, Junior, went prospecting. The thermometer showed 100 degrees when he left home that morning. Climbing hills was hot work, but by night the discovery of a good outcrop made it all worthwhile. His Amelia Mine was worked for the next 17 years and then sold for $10,000.

John Kelly financed Hamp's prospecting in the Rand area for a share in whatever he discovered. The Big Silver fully rewarded Kelly for his belief that Hamp was a prospector worth backing. After the silver strike, Hamp stayed to see it develop into a profitable enterprise. One year and a half after he knocked the hunk of blue-gray rock off of an odd outcrop, he sold out his share for a reputed $50,000 and moved to Los Angeles so that his son and daughter could obtain an education. Hamp, Senior, was 82 when he passed away in 1899, and Hamp, Junior, was 90 when he died in 1961.

The silver boom brought miners, prospectors, and the camp followers. Saloons opened and the girls arrived. Living quarters, usually at a premium in a new camp, were not so critical here; empty buildings from the now defunct Atolia could be moved in, and there were hotels, rooming houses, and even a few empty houses in nearby Randsburg and Johannesburg. An unusual situation developed, however, as mining families moved their houses to be close to the men's work at the Kelly. People who had tried to run the "fancy women" out of Randsburg, Johannesburg, and Atolia were now happy to have a place to live near the mine and paid no attention to the fact that next door was a house where those same girls continued to carry on their trade.

George Binko arrived during the silver boom via train to Johannesburg and stage to Randsburg, where he stayed at the Houser Hotel. The diggings at Atolia were closing, the Yellow Aster was being worked mainly by leasors, and the Kelly was just starting. Binko couldn't see why everyone was excited over mining; the towns seemed dull, completely lacking the glamor he had expected, but since he was here he would look the area over before leaving.

"I walked to Saltdale, about 18 miles, then on to Cantil, another 5 miles, and back to Saltdale. The manager was agreeable to my staying the night at Saltdale and put me in a cabin with two of his workmen. The next morning there was only one man and one cot. I asked what had happened to the other fellow and was told that my snoring had run him out. He took his cot and moved out under the smoke trees where he couldn't hear me."

An Irishman and his wife gave Binko a ride back to Randsburg. That too was an experience because it took all afternoon to make the trip and 10 gallons of water, as they stopped often to cool the engine and add more water. When they finally reached Randsburg, Binko went to the barbershop for a bath, changed into a light summer suit, and with the Irish couple attended the Saint Patrick's Day dance.

In April he went to work at the Kelly, bought a house in Atolia, and had Morris Atkinson and son Bill move it to Inn City, where Bill Foster, owner of the ground, was giving a free lot to every man who would settle there. His was the third house to be moved in from Atolia. He worked at getting it settled in place until after dark. There was no restaurant, and the Kelly boarding house was filled; so he built a fire in his yard and, using a shovel for a pan, fried bacon, eggs, and bread.

Confusion surrounds the naming of the early

settlements in the vicinity of the Big Silver. There was Osdick siding with tents and cabins, another group called Hampton, and still another referred to as Inn City. The last is thought to have been named by a coal delivery man as miners left notes for him and their friends saying they were in or out. The thing became a joke and each miner tried to outdo the other. Signs read Do Drop In, Come On In, In and Out, Never In, Out Not In, Ah Lives In, Seldom In. With the coming of saloons, gambling, and crib girls, the settlement was referred to as Inn City, Sin City, or Gin City.

Just a short distance away, a struggle developed between Pete Osdick and the rougher element. Pete hired one man to paint "Welcome to Osdick" signs daily and put them up. Each night the signs were torn down and "Welcome to Inn City" was put up in their place. The town of Osdick was recognized in 1921 and a post office by that name was opened the following year. The situation didn't change just because Pete had managed to get the post office. Mail continued to come to Hampton and Inn City until the Postal Department decided to end the confusion and in 1929 named it Red Mountain. Pete, one of the first residents in that area, never got over the blow, and to his last day had his mail addressed Osdick Town c/o Red Mountain Post Office.

While Osdick was the leading town, Frank Linstrom and his young son Ross would make trips to the desert to relax and prospect. They would rent a cabin at Osdick and hunt the hills to the east. Young Ross was in on the only find that appeared to be worth working. "Dad and men named Church, LePan, and Detmer were prospecting together a few times, and this one time they found what appeared to be a good stringer up near an old abandoned mine on Fremont Peak. Since it was late in the day when the discovery was made, they decided to return to town, gather up their camping equipment, lay in a stock of food, and quietly leave early the next morning so that claims could be staked before anyone got wind of their find. It was mutually agreed that not a word would be said to anyone about the prospect or their plans."

The following morning the men were ready to leave long before the town was awake, and young Linstrom, along with his father, supposed that no one knew of their movements. "You can imagine our surprise when we reached Fremont Peak at sunrise and found the mountain crawling with men. Dad later said there were about 100 men waiting for them; to a boy it looked more like 300. Many had set up a temporary camp, some were having breakfast, others pretended to be prospecting. All had picks and shovels. Most were just standing around waiting to see where our party was going to start laying out claims."

The elder Linstrom muttered something about men with big mouths and departed for town where he found out that one in the group had indulged in drinks at a local bar and had confided in the bartender in a voice loud enough for the card players across the room to realize a gold strike might be in the making.

"Funny the things a boy remembers," Ross Linstrom remarked, "to me Randsburg was a Chinaman sitting under a lean-to along Butte Ave. The Oriental's long, full sleeves and little black cap were as fascinating as the small containers of gold he bought and sold, storing them out of sight in his great sleeves. Johannesburg left no impression, but Osdick and the Kelly Mine meant big wagons with gray-blue ore piled high and men riding shotgun on every load."

Riding shotgun on the ore wagons had become necessary at the Kelly when the richness of the ore tempted miners to excessive high-grading. To many, the act was justified as a philanthropic means of relieving the owners of part of their income tax burden. Miners became quite creative in their methods of high-grading; one rather large gentleman tied his long underwear shut at the ankles and filled them with ore from an extra rich silver ledge. Coming up out of the shaft at the end of his day, the fellow lost his balance while getting out of the ore bucket and fell flat. The shift boss was understanding, and as he helped the burdened miner to his feet, quietly remarked, "Don't load up so heavy next time!"

A more ingenious method of making away with

rich ore developed well away from the mine. The newly milled concentrates were loaded in railroad cars for shipment. It was a simple matter to crawl under the cars that waited overnight for the next train out. Holes were drilled in the bottom of the cars and a temporary plug put in the opening. As the shipment pulled out, the vibrations of the train would soon loosen the plug and the rich ore poured out between the tracks. A person not associated in any way with the mine operation would gather the concentrates in safety, well out of sight of the town.

Max Hess, who came to the Rand area in 1919, was among the first miners to work ruby silver at the Kelly, of which the first 5 tons ran $8,000 per ton. Material of such value was sure to tempt. Max recalled the time Ed Grady was leasing a shaft on the Kelly; one of his men became too greedy and looked heavier than he should. Grady followed him to his car. "You've some high-grade in there," Grady remarked as the fellow crawled into the vehicle. Without hesitating the fellow agreed. "What will you take for it?" Grady asked. The fellow made a quick offer of $750. Grady bought his own ore for that price and still made a good profit.

Finally, high-grading became so prevalent that the management began to check lunchpails and bulges in clothing and made the men turn down their pants cuffs as they went off a shift. Two men were put on the wagons for the short haul from the mine to the mill. The ore wagons with a 10-foot long, 4-foot wide, and 1-foot deep box had a seat for the driver and an armed guard; at one time it was necessary to have a man riding shotgun on the back of the wagon or on horseback following the wagon.

Movie producers found the atmosphere around the Kelly ideal for some of their scenes and shot reel after reel of the men, wagons, and mill operation, giving each person that appeared in a scene a few dollars in appreciation. One mine strike scene became overrealistic when young Dutch Bennett took to heart a small role he had been assigned and actually knocked out three extras before the director could stop him and explain that it was only supposed to look real.

The silver situation as of July 1920 was given by John C. Wray, Randsburg correspondent for the *Barstow Printer*:

> The Kelly Mine is shipping five cars of high grade ore this week and while still short handed, as compared to three months ago, work apparently is progressing smoothly. The speculative market is on the up grade. The mint is purchasing silver under the provisions of the Pittman Act, which fixes a mandatory price of $1.00 per ounce for native mined and smelted metal. Since June 15, 3,200,000 ounces have been purchased, 1,700,000 ounces of which was for delivery at the mint in San Francisco.
>
> As was predicted in this column a month ago, speculation in silver bullion by an English-American syndicate is the inside reason for the sudden drop in London and New York of more than 30 points in 90 days. That the open market price recovered 23 points per oz. in less than 30 days, tells the story of syndicate speculation in no unmistakable manner. The lambs received a record breaking shearing, and the Pittman Act is all that saved the American miners from being victimized down to the hide and tallow.

Though the Kelly, the Big Silver of the Rand district, had an exceptionally good record for the extent of its underground workings and the number of men employed, the State Mine Inspector shut it down in 1929. There followed a series of shutdowns and six changes in ownership, the last being the transaction between Frank "Pop" Royer and Mines Exploration, present owners.

In all the years of its operation the big mine had only two fatalities, an almost unheard-of record. Carl Johnson died in a cave-in, in May 1934. In 1935 William Peterson, Warren Bob, and Jack Laird were overcome by fumes when testing a new dynamite. Bob managed to crawl out and summon help before the gas proved fatal. In 1941, "Tuffy" Landon was smashed to bits when he was caught between timbers in an unexplainable accident, considering his long experience in mine shafts.

Fred Weimholt recalled an "almost" mine catastrophe in 1939. He and other men on the night shift were working at the 1,100-foot level. When it was time for them to come to the surface, there was no response to their signal. For the next hour they kept signaling, but there was no recognition from the hoist man. Finally they decided to climb the dangerous steep ladders, each man realizing that a misstep by any one of them could send them all to the bottom. After the long exhausting climb they found no one on duty at the surface. Vic Johnson, the hoist man, had had a few drinks and gone home to sleep it off.

Rabbit hunting with the 12-gauge pump gun was a real sport, taking patience and skill. 1923.

(Author's collection)

The Lively Twenties

18

Discoveries of large, rich ore bodies can be covered in a few cold facts; it is the people involved that give life, action, and interest to the discovery. When barren, seemingly worthless acres are covered with a mine camp, almost overnight, there is bound to be some similarity to other camps where those same men had lived; yet each camp had its own personality. The camps, Hampton, Inn City, and Osdick, each valued its place in the mining world, and though only a few rods apart were divided in loyalty, strife for leadership, and social status. Combined as Red Mountain they became the last of the frontier mining camps, classed as one of the wildest, "wide open" boom towns of mining history.

Red Mountain not only provided relaxation and entertainment for the miners of the Rand area but was also the center of activity for the men working at Trona. The American Potash Company looked upon Red Mountain as a desirable neighbor that kept the undesirable out of their young community, yet cut their labor turnover by providing entertainment for employees who would otherwise leave the isolated area for the glamor of the city.

It was the Prohibition period, but authorities usually looked the other way and liquor ran freely; prostitution was also frowned upon in other parts of the state, but in Red Mountain both liquor and women were considered a necessity of the time and place. Saturday nights were especially lively, and many a Los Angeles resident drove the 150 miles to take part in the activities or just to enjoy the unbelievable sight of 30 saloons running wide open, with the post office being the only place in town where one could not buy a drink. There was no pretense of cover up; hotels, saloons, cribs, and silver were all closely associated.

The hotels had busy bars and gambling tables; the girls' quarters were upstairs or just outside; the atmosphere spoke of the passing of the frontier mining town, the last fling at playing to luck, laughter, love, and liquor. The widely known spots were The Shamrock, The Monkey House, The Northern, Little Eva's, The Annex, The Owl, The Pacific, and later the Red Onion and the Silver Dollar.

Musicians and performers were always welcome in the saloons as a means of drawing trade from the competition. Anyone who could play a fiddle, blow a mouth harp, or squeeze an accordion was favored with all he could drink during week nights. A good day's pay often accumulated in his hat if he was especially talented. The larger places usually hired musical groups for Saturday night. The "Tungsten Kid," a square-built fellow who had carried supplies into the Klondike, was one of the leading local entertainers; whatever saloon he chose to favor at night was sure to have a big crowd. The Kid danced, and the more he drank, the better he danced.

Those who still recall Red Mountain's wild days agree that most of the girls were good looking, clean, and with few exceptions enjoyable company. "A few were real devils, but they didn't stay around long." Those on the line usually went by one easy-to-remember name; where they came from or what their real names were was never questioned. That was their business alone.

The more popular girls were Lois, a bit stocky but a neat dresser; Kathy, a tall redhead; and Jerry at the old Silver Dollar, who was short, dark, and exceptionally pretty. Tex at the Owl had large expressive hazel eyes and lived up to her name with a heavy Texas drawl. Carmen was short, dark, a bit on the heavy side, but very good looking. Rita, a Latin, was petite and charmingly emotional. One of the many by the name of Rose was a real beauty and always popular in spite of the fact that most of the fellows knew she was especially clever at lifting wallets. Indian May and Cokie Joe were both good looking Indian girls; the latter got her name because she always ordered coke and gin.

Eva's place was famous for its library just off the bar; Irene's was a friendlier but not so popular house. Sugar Pie was the strictest with her girls, but she herself was good company, fun to talk to, and a wonderful dancer. Red Mountain Hattie, a buxom madam, mothered her girls, was selective of the men who came to her establishment, and enjoyed a nip with them while getting acquainted and deciding whether the gentleman was a fit customer for her house. Hattie's overindulgence was blamed when her Ford overturned near Lancaster on a return trip from Los Angeles. At the moment she had little cash on her, and Picket Brothers Garage was reluctant to do repair work on credit. However, when Hattie identified herself and put up a massive diamond ring for security, the repair job moved ahead without further delay.

As one early-day bartender recalled, "Come to think of it, the girls were mostly brunettes, a few redheads—and not a blonde among them. Most didn't stay long, they moved on always hoping for better money or better living conditions. A few stayed, married, and made good wives."

Shootings were frequent, but the girls of the night were seldom openly involved. Tall, willowy "Arkansaw" was one of the indirect causes of gunplay. More aggressive than most of the girls, Arkansaw had a number of older miners as regular patrons. One of the miners always treated all the girls to drinks when he hit a pay streak, and Arkansaw was never able to claim his attentions entirely for herself. Shortly after the miner became pumper at Mountain Wells, he went to Los Angeles for a few days; when he again appeared in his favorite saloon, Arkansaw chided him for having drinks and not treating a girl in the place. The old miner answered, "I don't have to treat you whores any longer; I've got my own woman now."

A stranger sitting farther down the bar remarked that the slander was unnecessary and an apology was due; when no apology was offered, the miner was shot on the spot. Though Red Mountain was in San Bernardino County, there was no move to handle the case; finally due to local pressures, the sheriff in Mojave took over. A search of the hotels and cribs located the murderer hiding behind clothes hanging on the wall of one of the girls' rooms. Investigation revealed that the accused was hiding out in Red Mountain after having killed a man aboard a ship tied up in Los Angeles harbor.

Because law officers were reluctant to become involved in Red Mountain problems, many incidents were swept under the rug and forgotten. The "Mid-Way Shooting," following an argument in one of the bars, was like many others; it happened near the Kern County-San Bernardino County line and neither side wanted to touch it. More exciting shootings soon pushed it into the background and public interest centered on more recent happenings.

Liquor raids on Red Mountain were relatively few; a number of laws were openly disregarded. Fines were light and imprisonment was seldom considered necessary. Sanitation was the only real restriction adhered to; American Potash recognized the workman's need for girls, whiskey, and fun but retained Dr. Denton and later Dr. T. Drummond to make regular examinations of the girls for the protection of their employees.

The girls of the night kept to themselves, seldom having any contact with the other women of the community. Even the most strait-laced admitted that most of the girls were well behaved in public, and as one wife of a Randsburg citizen remarked, "They were a high class girl for the type they were."

Considering the small, poorly furnished quarters in which they lived, the lack of adequate water or sanitary conveniences, it is commendable that they were able to maintain personal cleanliness, appear well dressed, and remain entertaining.

The live music of the bars and dance halls was usually provided by a group of young fellows from Los Angeles. The fellows would pile into cars and start for the Rand area early Saturday morning, arriving in time to play for the night crowd. Travel was slow over washboard roads, and flat tires were an accepted part of the trip. The pay was considered well worth the effort, each musician receiving about $15.00 and a night's lodging. As one drummer who frequently played at the dances in Randsburg and Red Mountain remarked, "There was always a good crowd and good spenders, and competition among the girls usually added up to an evening's entertainment."

Harold Berry and a friend came through Red Mountain looking for work in 1921 and found that, "In the daytime the barbershop and attached bath was the busiest place in town, but at night all the activity was in the bars where the girls were good looking and bootleg whiskey sold for $2.50 a pint; the whiskey was so strong that you got full value for your dollar."

The two young fellows were hired by the Yellow Aster Mine for a short-term job, timbering in one of the tunnels. Pay was $1.75 for a ten-hour day and use of a one room corrugated metal shack in Red Mountain. It was early spring with cold rains turning the dirt street into a mud puddle that soon flowed over the board walks. Berry added, "You knew you were on the board walk if the mud wasn't too deep and you hit a firm surface. Since you couldn't see the walk it was easy to step off the boards and find yourself wallowing in deep mud that seemed to have no bottom. It was impossible to be clean with mud everywhere and water at 25¢ a gallon; at $1.75 a day earnings you couldn't spend much on baths."

Michel F. Jackman, a seasoned miner from the camps of Colorado and Nevada found Red Mountain, "running wide open and wilder than other camps of more fame." He added, "On the other hand, Rand area had fewer mine accidents than most mining areas; but men injured and killed in Red Mountain drunken brawls more than made up for it."

Within one week there was the blacksmith found dead in bed after a saloon argument, a barber of one day was shot as he left his business, another fellow was found shot to death in jail, and a deputy was ambushed while he was taking a troublemaker out of a bar. It is no wonder Jackman and other miners familiar with the rough life of mining camps found Red Mountain "a bit worse than most."

Shortly after Jackman arrived in the area, he

witnessed a Ku Klux Klan cross burning back of Red Mountain in protest against the saloons. Signs reading "Beware of the Invisible Government" were tacked on poles along the street. To compete with Red Mountain entertainment the K.K.K. put on a free Saturday night dance at Johannesburg in a building now used by the Four Square Church. The only requirement for admittance to the dance was to satisfy the guards at the door, who took a whiff of each arrival's breath and refused entrance to anyone with a trace of liquor on his breath.

Not all the life in the Rand area was as lively as that in Red Mountain. Some soldiers returned to settle into the serious business of carving out their own niche in the community. James B. Nosser, Jr., son of one of Rand's earliest miners, came home to open the Hampton Garage, Blacksmith Shop, and Assay Office in 1922. The blacksmithing was more important at the time than was the garage, for few cars traveled the rutted, dirt roads through the area. His was the first business to offer service to the traveler. In fact he received commendation from the Auto Club for serving the largest area in Southern California; his was the only garage in the area from Mojave on the west, Barstow on the east, Lone Pine to the north, and Shoshone over toward Death Valley.

Having the garage brought Nosser in contact with investors who drove as far as this outpost and then asked Nosser to guide them or take them into the isolated mining areas. One day a Long Beach banker wanted to see some mines and especially a rattlesnake. "The mines were easily supplied, but the rattlesnake was something else, since they don't usually put on a command performance," Nosser related. Out near Granite Wells, in spite of the local man's advice, the banker barged up a steep bank only to suddenly start running downhill backward with a snake gracefully gliding along toward him.

"That banker was in and out of trouble all day," Nosser recalled." He ended up by scooping nuggets off a riffle board up in Goler just as the miners were about to clean up their day's run. Nosser quickly saved the banker by explaining his ignorance of mining courtesy. "He could have suffered a beating at the very least; no one touches another man's riffle board. In this case the nuggets were fair size, some as large as kernels of corn and the day's run netted near $900. No one would appreciate a stranger dipping in for a few nuggets!"

Shortly after the banker episode, Nosser was driving near Garlock, when a new 1925 Packard came roaring toward him and stopped in a cloud of dust while the driver asked directions to Nugget Gulch. Along with the directions Nosser explained that the road was so bad that a passenger car couldn't make it. Nevertheless the fellow was so intent on finding the gulch that he paid no attention to the warning. He showed Nosser two 3-gallon pails and asked if they would hold all the nuggets he could pick up. Without waiting for a reply, he drove off in a swirl of dust. When Nosser got back to the garage, he sent a truck down to rescue the man, telling the driver exactly where in the sandy wash he would find the Packard and the man who was going to fill pails with gold nuggets.

The same year that Nosser was starting his blacksmith and garage business another veteran was starting a mining operation on the edge of the old Summit Diggings. Clarence Wadsworth had married an Army nurse and brought his bride to a dugout home 10 miles from Randsburg. Though she had seen primitive living while in the service, the hole in the side of the mountain came as a shock. She had experienced lack of food at times during the war when the hospital had been shut off from supplies, but there had been work to keep the mind and hands occupied. In the dugout when there was no gold to buy food, there was nothing to do, no place to go, and the days hung heavy. When the mining went well, there would be a trip to town, some fun at Red Mountain, shopping for food, and then back "to the hole in the ground. There never seemed to be enough gold to build something better or get us out of there," Ruth Wadsworth recalled. Yet she stayed; and Wadsworth did his best to make the place more comfortable and still make a living mining. When he developed silicosis, she nursed him through a long illness; and then as a widow 24 years after she came to the mining country, Ruth Wadsworth left in search of old friends and the city life. People had changed; the world she had known was strange; and the previously unrecognized charm of the mines, the mountains, and the people brought her back to Randsburg.

The 1920s in the Rand area were years of contrasts; while Red Mountain whooped it up, miners like Wadsworth struggled to be independent of a mining company and make their own way, and the settlers around Cantil fought the sand that threatened to wipe out their years of work.

Five years after the Red Rock school was built by the people of Cantil it was becoming almost useless; sand settled in the attic, making the building a safety hazard for the students, and the exterior was engulfed in sand dunes after each wind storm. To save the building it was moved farther to the west to a new location on land donated by A. J. Crookshank, where the school yard and the new building still can be seen.

Moving the school was a remarkable feat considering the sand and the horsepower and the unique equipment available. Men worked two days shoveling sand out of the attic to relieve strain and weight. Under the supervision of Edward Roe, an experienced house mover, men of the community worked days battling the sand as they jacked the building up. Three wagons were stripped down and shortened up; space between the wheels was filled with railroad ties, with long timbers placed on top to form a supporting frame for the building. One wagon was at the front; the other two were at the back. A 20 mule and horse team from the Van Winkle and Crookshank ranches was brought together, and Pete Cassou, one of the district's earliest pioneers, drove the long team of animals that had never worked together before, an accomplishment that the freighters would have admired.

Slowly the building moved down Norton Road, over sand dunes, and through ruts. Men volunteered their work, materials, and animals. It was a community working together to accomplish the near impossible.

Just a few miles away, Red Rock Canyon was becoming known to the leisure class who could afford the adventure of travel. An increasing number of cars made their way through the colorful canyon, although it was still a dangerous trip, with miles of desert, infrequent watering places, and few places to stay. For the venturesome the Department of the Interior made a study of existing water sources, listing water and supplies at Red Rock Canyon, Koehn Springs, and Saltdale; water, gas, supplies, and emergency housing at Garlock; water, gas, supplies, hotels, and garage service at Randsburg and Johannesburg; water, gas, supplies, and emergency housing at Red Mountain. Atolia was listed as closed, with only water obtainable from the watchman.

The automobile also brought in a new wave of the lawless, who found it easier to work an area and depart when they had pushed their luck to the breaking point. People became uneasy when strangers hung around without apparent means of livelihood or an attempt to find work. Sue Reeves, Railroad Express agent in Randsburg, had been warned against possible thievery; whenever shady-looking characters lingered near her office, she made a point of leaving a light on all night and the door of the safe open. Usually every valuable thing went out on the night train, and there was little to worry about, but one time a gold brick was brought in for shipment too late to make the outgoing train. Sue had noticed a fellow hanging around town for a couple of days, often standing across from the Express office. That night she told her husband Gene, manager of the Rand Mercantile, that she was worried about the gold brick still in the safe. "That night," Mrs. Reeves later revealed, "we slept with a gun and a gold brick in our bed." The next morning everyone was talking about the 16 places in town that had been robbed during the night. The Express office was untouched, with the light still burning and the safe door wide open.

Street traffic in Randsburg.

19

The Good and the Not So Good

On May 17, 1923, the *Bakersfield Californian* carried the headline, "Kern Man Held in Bomb Attempt." The article stated:

> Charles Koehn, Kern County 'Desert Rat' by his own description, a resident of Saltdale on the Mojave desert, is being held in the Fresno jail in connection with an alleged attempt to bomb the home of Superior Judge Campbell E. Beaumont early today, according to a special dispatch to the *Californian*.
>
> The bomb, consisting of 2 parts, one containing the explosive and the other gasoline, was found by the jurist when he opened his front door to get a newspaper. The fuse had extinguished itself when less than a foot remained.
>
> Koehn was arrested by the police as a suspicious character prior to discovery of the bomb as he ran from an alley two blocks away from the Beaumont home. In his automobile was found a quantity of fuse which matched that attached to the bomb. While Koehn denies implication and said he was lost, the officers found torn portions of a Los Angeles newspaper in his car, which matched a piece wrapped around the top of the internal machine.
>
> Both explosive and gasoline were contained in discarded coffee cans, blackened and rusted with age.
>
> When Koehn was led before him, Judge Beaumont recognized him as a defendant in a $50,000 damage suit in which the jurist sat in Kern County a few weeks ago.
>
> The Judge said today 'I can only conceive what Koehn, if he is the guilty person, could have against me as the case in which he is involved as defendant has not been settled.'

The following day more was added to the story: if Koehn was guilty, two charges, attempt to commit murder and malicious use of explosives, would carry a maximum sentence of life imprisonment.

The bomb was home-made from a pound coffee can filled with putty, sand, and nitroglycerin. The caps were not the type usually used in mining but were the heavy-duty kind used by oil well companies in "shooting" a well. On top of the coffee can bomb was another can full of gasoline, which would have assured total destruction of the judge's home had the fuse burned to the can.

Although he continued to deny knowing anything about the bomb or where the judge lived, Koehn was arraigned in court 2 days after the bomb attempt, and his Rand area friends were attempting to raise $25,000 bail.

Though Charles Koehn had been known as the "Wild Dutchman" in the late 1890s when he would get riled up over prospectors stretching their credit too far, those who knew him best couldn't imagine his becoming so upset over a court case that he would consider doing harm to the judge. As one Cantil resident remarked as he added a share to the collection for Koehn's bail: "I don't believe he did it, and we got to get him out pretty soon or he's going to miss the big doings over at Cantil when we fix up the burial grounds May 30th." Another was quoted as saying, "Koehn promised to let us eat under his trees and said he would make a speech that day."

Koehn was liked by some, feared by some, and heartily disliked by others. Some thought he was a sharp old man, always fighting someone or something. Those who liked him had come to understand his heavy accent and enjoy his brand of humor, frequently at his own expense. He never failed to enjoy telling about the cache he had under a hen's nest right where his wife daily gathered eggs. Since his wife wanted no liquor on the place, Koehn would wait until a friend arrived and then they would wander down by the chicken house. With no one around to notice, "a small nip" would be produced from under a nest. When the guest was gone, Koehn would remark to his wife that the visitor had been most thoughtful in sharing a drink, which had helped to relieve his shingles.

Even in San Quentin the little Dutchman had his laugh. Koehn would write a friend telling the near location of a lost mine; then he would write another telling him the joke played on the first one, saying that the lost mine was really where he had marked the map. Knowing that not one of them could resist a look at the suggested location, he had men running all over the country hunting a lost gold mine in the vicinity of the *X* on a map. Even when he would kid them about it in his next letter, most still believed that there was such a mine and that Koehn had given one of them the right location.

The rantings against those he felt had stacked the evidence against him, and the laughter with those he knew as friends came to a halt on August 7, 1938. A few days before he was due for release, Koehn died in prison.

Whether or not Koehn had some knowledge of a lost mine is still a point of argument among the few remaining miners who knew him or were party to his letter writing. But lost mines never fail to have a following, and the Rand area had its share reported through the years. Judge James B. Nosser, Jr., told of finding a rich gold deposit that later became known as Nosser's lost gold mine. He had been prospecting along the side of Red Mountain when he suddenly came upon pieces of float spotted with gold. Since it was growing dark, he noted the location and returned the following morning; right off the surface he filled a number of ore sacks with rock, heavy in gold. By evening he could find no more, although he dug in several places and hiked over an extended area around the first find. He decided that it was a "blow out" and that he had all there was at that location.

Nosser sold his ore the following day amid a flurry of interest, and for days thereafter every move he made was closely watched. As time went on and he failed to bring in more of the rich ore, the rumor started that he could no longer find the deposit, and a lost gold mine was born.

Years later he returned to the spot and diligently searched for any scrap he could have missed or that might have eroded out, but he found nothing. It was his theory that many lost mines were never lost; they had been only a small pocket of rich ore that the miner extracted at one time, and there was no need to return.

The Lost Wishbone Mine was something else. A

miner came in from the El Paso Mountains with a small ore sample to be assayed. The assayer, busy with runs at the Red Dog mill, told the stranger that the sample was too small to give an accurate assay and to come back later with more material. The lone rock lay on the shelf waiting, but the fellow never returned. Eventually the assayer cleared the shelf, tossed the rock into the waste basket, and dumped the lot on a rubbish fire at the back of his house. Later in the day he checked the ashes to be sure there was no fire left to endanger his house if a wind arose. As the man kicked through the ashes, he was astonished to see what had been a primitive assay in action; gold had melted out of the discarded rock and adhered to a charred chicken bone. The excited assayer tried to recall what the prospector had said, but realized that his own abrupt manner had not encouraged conversation. Where that rock came from there is a rich gold deposit, still lost as far as anyone knows.

Not all lost mines were gold. A father and son prospecting out of Death Valley found a ledge of rich silver somewhere between Searles Station and Red Mountain. They worked out as much as their burro could carry and brought it into Johannesburg to exchange for groceries and cash. As usual, word traveled fast, and when they left for the diggings the following day, they had lots of company. A two-day search turned up nothing; they didn't recognize a single familiar landmark. By the time the disappointed miners returned to Johannesburg and the two men had gone on to Death Valley, the Lost Swede Mine was an established fact in spite of news that the assayed silver was suspiciously similar to that being brought out of the Kelly Mine at Red Mountain.

Down in the valley, at Garlock, other lost mines were in the making. Martin Engel, Cantil postmaster and last of the fire assayers, was a young prospector living in Garlock at the time one mine remained a mystery. On a hot summer night he and a number of other fellows were sitting outside Miller's Bar that had been converted into a rooming house. Engel thought he saw a light appear and disappear up on the mountainside to the north. He waited for someone else to see it and make a comment. No one said a word; so he thought his eyes were playing tricks, or that he had been on the desert so long he was beginning to "see" things. Finally Engel ventured to ask if there was a light on the mountain. "Yeah, it's a guy that digs where the spirits tell him, and they usually aren't around except at night," was the reply.

Engel then heard about the gold the fellow sold: "So he had to be working it out somewhere, but when anyone attempted to follow him, the spirits went away and there was no digging." By day many searched the area where they believed the light had appeared the night before but found no sign of pick or shovel work. That mine wasn't exactly lost, but where the fellow with the lantern got his gold was lost as far as the rest of the prospectors were concerned.

Aunt Emma Conners, who lived at the eastern edge of Garlock, was party to another lost mine. Famed throughout the mining country for her delicious cheese, Aunt Em daily grazed her goats along the foothills of the El Paso Mountains behind her home. The children of the town loved to go with her. While the goats grazed, Aunt Em would tell exciting stories of her childhood trip across the country in a covered wagon, and at noon she would provide a delightful lunch, goat cheese included. When the animals strayed, the children earned their stories and lunch by chasing after them, throwing small rocks to drive them back to the herd.

One evening when the group returned to Garlock, one of the Hickie twins had a rock in her apron pocket that had been too pretty to toss at the goats. When she showed it to her father, he became very excited because gold was cropping out all over it. Hickie crushed and panned the rock that night and was almost tempted to wake the girls to find out where they had been that day. The next morning when asked where she had picked up her pretty rock, the child swung her arm in a wide arch from east to west. The goats had wandered far that day, and she had picked up many rocks. Every miner in town took a turn at questioning the twins, plying them with the best candy the local store had to offer, but to no avail. Prospectors searched the foothills in vain for another chunk of the float that would give them a key to the location of a possible ledge of rich ore under their feet or higher up on the mountain.

Old timers still think that out there somewhere, looking down on the townsite is the Twins' Lost Gold Mine and that someday its finder will be rich.

Just over the ridge to the west is another lost mine worth looking for; it was lost during the depression days when more and more miners were making a living out of gold deposits that had been bypassed by earlier prospectors.

Old Joe had been working the draws and canyons of the El Pasos for some time before the night he rushed into a saloon and dropped a bag of high grade on the bar. Everyone became excited, and Joe used up his gold samples on drinks. As the night wore on, men wandered home, planning to follow Old Joe in the morning and stake some claims for themselves. But the barroom gals got a jump on the men; they talked the old prospector into taking them up to his discovery and blowing up the mound that he was sure was nearly solid gold. All the girls wanted was a souvenir chunk of his gold hill, and Joe agreed that he would never miss a few pieces of ore.

The sleeping residents of Garlock were shaken out of their beds in the wee hours of the morning by a thundering blast. Lanterns appeared, and men ran back and forth shouting at each other and searching for the source of the explosion. It was not until daylight that they found out that Old Joe and the girls had dynamited the top off his gold mound expecting to pick up chunks of rich ore in the light of their lanterns. As prospectors of that period sum it up; "Old Joe either shattered his gold into bits, or in his liquored state, blew up the wrong hill." No gold was found; Old Joe's mine was lost.

Lost mines, like the turn of a card at the green felt table, grasp man's imagination and hopes. The next search, or the next card hand could turn up a winner.

The lean years of the depression may have lent extra emphasis to the hope of finding the wealth of a lost mine or a rumored rich vein. The Rand area as elsewhere felt the financial change; the mining boom atmosphere was lacking. Prospecting involved more and more men; as in the earliest mining days, even a little gold kept the wolf on the other side of the door.

Goler Canyon had crawled with prospectors in 1893, and though more gold was found elsewhere in the years to follow, the canyon continued to be worked off and on. When the James Argyles moved there in 1929, there was no one else living or working the once rich placer area, but within days other families started moving in. Each new arrival hoped to find enough gold to provide food for the family, which was more than they had been doing in the city.

Argyle located some good claims, but there were lean days before the gold showed up in sufficient quantities to pay for food. Mrs. Argyle told of a three-day period when all they had was a little oatmeal that she rationed out for their children; she and her husband worked their claim with nothing more than water in their stomachs. However, the crisis passed as their work began to pay off, and though there was no surplus, they would gladly have shared with a family farther up the canyon if they had known how much they needed help. The family with six children were living in a dugout, had eaten their dogs, and were trying to get value out of the desert tortoises by the time the Argyles became aware of their plight.

Even when the Argyles found gold in sufficient quantities to ease their financial worries, there were still the many rattlesnakes and claim jumpers to deal with. In spite of parental worry and warning, the children would play hide and seek in the brush where the snakes were plentiful. One of the Argyle boys was missing seven hours one day; every man in the canyon went on a search of the old mines, sure that he had fallen in an abandoned shaft. When they found him, he was being held prisoner in the entrance to a tunnel by a rattlesnake coiled just outside.

The claim jumpers at times were as dangerous as the sidewinder snakes, and Bruce "Curley" Carroll was one of the worst. Living in Sand

Canyon, a branch off the main Goler Canyon, Curley kept an eye on all the mining operations above and below him. When a claim was not being worked at the moment, he would sell it to some unsuspecting stranger. He sold some of the Argyle claims, and was told to stay away from their property from then on. But one day when Argyle was away, Curley came up the canyon, drunker and nastier than usual and ordered Mrs. Argyle and the children out of the canyon. In his raving he threatened to shoot them all and added that he had shot his teacher back East and no one had caught him yet for that one.

When Argyle arrived home and heard of the episode, he decided to do some gun waving himself and marched over to straighten Curley out on a few matters. Some days later the family was sitting in their two-room cabin when suddenly they heard shots out back. Supposing that Curley was on another rampage, Argyle rushed out the back door but found no one; instead he discovered that the newly bottled homemade root beer had fermented in the desert heat and was blowing bottles all over the dugout back of the house where it had been stored in the coolest place available.

Not only did the miners in Goler have trouble with Curley, but the bartenders also found him difficult and "downright mean the more he drank." Swanee Moore, bartender in The Red Dog and later the My Office of Randsburg, had tried to keep Curley out of trouble a number of times when he started irritating other customers. One night when Curley came in saying that he had found a seven-pound nugget and expected special privileges, no one paid attention to his blowing. But when Curley went behind the bar to serve himself the liquor he thought his find deserved, Moore had reached the limit of endurance. In a flash the bartender lifted Curley's old felt hat, slapped his face with such vigor that Curley's false teeth went sailing through the air, and yelled, "Now you get out of here and back down to your diggings."

As more families moved into Goler Canyon and the gold paid off, the residents saw their way clear to raise money for a building; with donations and volunteer help they put up a schoolhouse that was also a social center and church. By the time Hank Walker arrived in 1932, there was quite a community spirit evident, and though everyone worked hard and pay streaks were sometimes far apart, life had its pleasant side.

Walker worked the Edith E. Claim, paying 20% of what he found for the lease. "I worked four winters," Walker explained, "and took out $2,500 in gold. I'm the fellow who laid out a full golf course in that wash—18 holes and only six good ones!"

Though the cream of the gold had been skimmed off by the turn of the century, Walker felt that there were still pockets to be found in the canyon, "The whole trouble is that a lot of hard work and time are involved between rewards," was his comment. He told of "Scotty," a good miner who couldn't leave liquor alone. Saturday nights usually found him at one of the mining towns enjoying the dance. One night when he was at the Cantil dance, drinking heavily and looking for a fight, no one would become involved. Finally he did get some trouble stirred up and pulled his .45, which slipped out of his hand and went spinning across the dance floor. There was a grand scramble to get the gun before Scotty knew what had happened.

Scotty finally realized that his welcome was wearing thin. To get back into the public's good graces he did some serious mining and worked right into a good pay streak. On the way from his camp to the workings he passed the Brady camp, and when word got around that Scotty was doing well, the Brady girl just happened to be out by the road every time he passed. Soon Scotty was dating the girl, and the drinking sprees were a thing of the past.

Suddenly the pay streak pinched out, Scotty's income decreased, and Old Brady ran him off. Angered by this turn of events, Scotty would level a few shots at the Brady house each day as be went by; in turn Brady had him arrested. Scotty left the country when he got out of jail, and a fellow known as Bersell took up Scotty's Lower Fine Gold Claim. Bersell was an expert miner, and the tunneling he did was a neat, safe job that in a

125-foot length and 35-foot width produced $18,000 in gold.

During that same period Jehovah George moved into the canyon; his partner at the start tagged him with the name as he was "dead set against hard liquor but felt that wine was most acceptable." The Saturday dance at the Goler schoolhouse was the social gathering of the week, and after he got acquainted, Jehovah George was always there. A couple of Mexicans, playing violin and accordion, were the favorite musicians. One Saturday night some fellows brought a jug of "40 rod," the term used for poor whiskey that could knock a man out before he went 40 rods. Some of the men decided it would be really funny to get some of the booze down Jehovah George. When they invited him to have some of their "real good wine," he tipped up the jug and took a long drink as he would have of wine and then proceeded to scare everyone by nearly strangling.

Happy Hague created another kind of excitement. Looking after the Chamberlin claims for what gold he could work out, Happy was just making enough for groceries when he decided to work out a curve in the stream bed that had gone untouched by earlier miners. It was a good decision, as gold appeared in every shovel of gravel tossed on the drywasher; then Happy hit a 23-ounce nugget and became so excited that he couldn't keep the secret, but rushed into Randsburg to share his discovery. Liquor and the dance hall girls had absorbed most of the value of the nugget by the time the news reached Chamberlin, who started a search for Happy, hoping to claim a share.

When Chamberlin eventually reached his claims in Goler, the sobering Happy was waiting with a 30/30; Chamberlin decided it would be wise to go for the sheriff. Happy guessed the man's plan and followed him out of the canyon. They drove their old cars at top speed down the rough, one-way road, throwing sand on every curve. Each hoped to reach the law and tell his side of the story before the other had a chance. His prior agreement with Happy gave Chamberlin no right to a share, and Happy was not easily intimidated, especially when the nugget was already spent.

Though nuggets were not found frequently, the majority of the residents in Goler made a fair living after their mining operations got under way. Few were aware of the less fortunate who were traveling the road past the mouth of their canyon. Dilapidated vehicles of all kinds, loaded with families and all their belongings, moved along the road through Randsburg or the other route through Garlock, en route west toward the San Joaquin Valley and possible jobs in the fields of that prosperous agricultural area. They had entered California through Death Valley to avoid the check stations where their only worldly goods might be taken away in an effort to discourage them from continuing their trip, and to help convince them that they should return to their own state.

Desert travelers.

Civilian Conservation Corps boys working to improve roads in Death Valley were frequently called upon to help those who had entered the valley without adequate water, food, or gas. The supplies provided in the valley would run out about the time they reached the Rand area. Aunt Em Conners, and Sarah "Granny" Slocum in Garlock were both known to help anyone, miner or transient, who needed a meal. Granny was sharp-spoken and sometimes cross, but no one left hungry, and usually there was a packed lunch "for the children," that would see the whole family through to the San Joaquin Valley. These job-seeking nonresidents seldom paused for long in the desert communities for there were no jobs for those without experience in mining. They were too desperate to take the time to learn mining and the long gamble on finding gold.

Bill Stevens panning gold in his dugout in Cave City, Last Chance Canyon, 1928. *(Author's collection)*

In 1931 "Quick Millions" being photographed in Red Rock Canyon. Spencer Tracy was leading man in movi depicting gangster period. *(Author's collection)*

20

Moonshine and Sunrise

During the depression years all the old placer fields of the Rand area were being reworked, and as in the early days, strangers arrived and departed with little notice. A new set of names began to appear on credit ledgers of the local stores, the voting register, and the gossip column in the local paper. Every one was trying desperately to eke out a living; a few found ways of adding a bit more to their earnings.

In Last Chance Canyon the Cudahy's Dutch Cleanser Mine was operating, with Mexicans making up most of the labor force. Between the company store and the old cars brought in to sell to the workers, the owners kept the Mexicans continuously in debt to them, completely eliminating the transient labor problem. Once, feeling the need of a little extra income, the Superintendent decided to cash in on the fact that some of the company owners were bringing their secretaries or other feminine friends for a weekend at the camp.

Soon his blackmailing project was running along so well that he decided to spend a part of his fortune on a weekend fling in Los Angeles. Not willing to find his own entertainment, he pressured a company official into providing an evening in the best company and the best night spots. In his newfound power the Superintendent was ripe for a set-up. The girl, the photos, and the witnesses ended his blackmailing and his job; a jail sentence was dangled before him; so he left the country.

Saltdale, a few miles to the east, was a company-owned settlement that provided jobs, living quarters, and utilities. The Long Beach Salt Company took over the Fremont and Consolidated in 1931, developing the brines of Koehn Dry Lake an underground salt lake, into a profitable operation. An assembly line of men was necessary to harvest the salt from the evaporative area through the milling process. The best harvesting months and highest need for workers were the hottest, with temperatures running to 110 degrees and up at the mill and much higher out on the drying beds where the white of the salt and the millions of crystal faces reflected the heat at greater intensity.

Most men were glad for a job at a mine or a mill, but others wanted quicker returns for their efforts and more excitement than the routine jobs offered. Bootlegging offered both. A moonshine project in an abandoned mine tunnel between Saltdale and the Cudahy Camp developed into a most lucrative business and caught the attention of the "Revenuers." As the officers converged upon the suspected still, the operators rushed to conceal the evidence by pouring gasoline over the sugar mash. The danger passed and as the men viewed the loss of the income, one suggested that they spread the mash mixture out so that there would be a greater evaporative surface in hopes that the gasoline would depart and they would not lose the whole run. People traveling the main road commented about an odor that hung over the area for miles, explaining to friends that, "It smells like gasoline and something."

Old Bill Stevens in Cave City, up in Last Chance Canyon could have steered the Revenue men to other locations that he had discovered as he prospected the El Paso Mountains, but Old Bill kept to himself. Rumor was that he had been a hired gunman in Montana, but none dared ask; the old double-barreled shotgun carried across the saddle of his mule and a .45 on his hip were enough to stop the curious.

Bootlegging, seen through the eyes of a Revenue man, was breaking the law, and it had to be stopped or at least discouraged. One officer stayed at the Cottage Hotel in Randsburg part of the winter of 1929 while trying to run down the source of booze that was flowing through the area and even supplying a Los Angeles clientele. With strangers coming and going continuously, the Revenuer was accepted as just another mine investor, salesman, or job seeker. By listening and taking part in conversations with the old-timers as well as new arrivals, he found enough leads on the main liquor source to pay well for his stay in the cold old hotel with an outdoor facility that commanded a waiting line early in the morning regardless of the icy wind beating at them.

It was a three months' job, but the Revenuers moved in on a big moonshine setup in a canyon near Tehachapi and crippled the main liquor supply coming into many communities between the source and Los Angeles, the big market for the illegal stuff.

In the Rand area, word usually traveled ahead of the "Feds," and by the time they arrived, everything looked legitimate. One incident was recalled by Michel Jackman; he and a long-time friend of the mining camps were living in Randsburg at the time a fellow was killed in a drunken brawl at a hotel just up the street. "Rosie, the hotel manager, came hammering on our door till she had us up and then asked if she could store some of her supplies in a room we weren't using. It seemed an odd time of night to be wanting to store supplies, and we didn't quite grasp what the emergency was at the hotel, but Rosie was welcome to the empty room. Those supplies turned out to be home-brew that she was afraid would be discovered when the police arrived to investigate the killing."

When the end of Prohibition was in view, John Norton, merchant and postmaster in Garlock, though badly crippled with arthritis, was determined to celebrate when the great day arrived. He had a small, temporary bar installed in his establishment and invited guests for the hour when liquor would again be legal. A guest book was placed at the end of the bar for each guest to sign. The list grew and the distinguished guests followed their signatures with M.D., Ph.D., Engineer, Chemist, etc.; when George Stroop arrived, he looked at the names preceding his and decided that he needed to add something more. His entry read: George Stroop, J.P. (just a prospector).

Shortly after that Norton moved his business to Cantil where he continued as postmaster until his wife took over. His grocery store supplied the ranch community and the travelers exploring the Red Rock and Last Chance Canyons. Local residents were especially appreciative of some of his methods of handling city slickers and hunters. Norton sold shells in his store and always appeared to be helpful to strangers inquiring about the best place to hunt by pointing out the flats near Cinco. When some one would kid him about sending the hunters where there was nothing, he would reply, "You don't think I'd send them to our best hunting grounds, do you? And maybe there isn't anything around Cinco, but there could be."

While moonshiners had been building one kind of reputation for the Rand area, another group had been just as busy building the opposite character. In 1929 a close-knit group at Cantil planned an Easter Service in Red Rock Canyon, where the eroded formations provided a natural temple. Bessie Arnet and Jenny Koehn decorated a wooden cross with the pale lavender swamp daisies growing at Kane Springs; others brought hymn books and Bibles. Though nearly everyone in the Cantil, Cinco, and Red Rock area turned out and participated in the scripture reading and songs, the little group and the cross that seemed so large to haul in were lost in the immense open amphitheater and maze of eroded spires.

The next year Carolyn Engel made huge paper lilies for the enlarged cross; everything was on a larger scale; and other communities were invited. For that sunrise service people came from Randsburg, Johannesburg, Goler, and Garlock to swell the attendance and add their voices to the singing. The huge wooden cross that had posed so many problems in being hauled up to the highest point was left standing for the following year, because now the originators of the Easter Service felt that there was a real desire for the observance to continue.

During the year vandals shoved the cross over the cliff and used it for firewood; so the whole difficult process was undertaken again the next spring. But the interest grew, and the third year people from as far away as Tehachapi attended. A long row of cars filled the background, and a good-sized crowd stood in a half circle near the minister and choir.

In the years that followed Martin Engel's Boy Scout Troop spent the Saturday before Easter, cleaning the canyon, and camped overnight to keep an eye on the cross. Others in the community got up hours before time for the sunrise service. Walter Tisch, Road Commissioner, built a huge bonfire and had gallons of coffee for those who stood in the cold morning dawn. Stover Lowe, of the aqueduct station in Jawbone Canyon would go about five miles to Ross Rogers' place for the organ and get it set up before anyone arrived. The women made preparations to serve a sunrise breakfast after the service.

Because there was no loudspeaker system and the people stood close to the pulpit so that all could hear, Lidia Atkinson's voice, too powerful for most community buildings, filled the canyon with melody and led the audience to put forth its best voice on the hymns.

First Easter service in Red Rock Canyon, 1929.
(Courtesy Martin Engel)

Old-timers who have attended and helped with the sunrise service through the years believe that the most impressive services were during World War II years when gas was rationed but people still drove great distances to participate. The church groups on the Navy installation at China Lake helped with the service, and the Navy furnished a portable speaker system. With the soft pinks, tans, and yellows of Red Rock's cliffs as background, the altar draped in deep blue velvet and holding the crosses of all the religions represented, was a beautiful, impressive sight.

One year Walter and Wally Bectel, owners of a store on the edge of Red Rock Canyon, built a cross to the style used in Biblical times. With rope and tackle it was slowly worked to a spot that seemed so difficult to reach that vandals would not bother, but in a short time it, too, was gone. Then Stover Lowe built a steel one following the dimensions of the original. Cement was laboriously bucketed up to an almost inaccessible spot and the massive metal cross set on a permanent base. The community believed that at long last their struggle to erect a cross had come to an end. But, alas, that summer, a film company sent a wrecking crew to cut it down because it interfered with a picture they were shooting. No effort to place a cross on a high canyon rim has since been attempted.

As the attendance at the services increased each year, many things were tried to improve the impressiveness of the occasion. One year a youth from Ridgecrest crawled up the cliff and stood out on a narrow eroded shelf to play a trumpet welcome to the rising sun. Everyone stood in awe, caught up with the inspiration of the boy, the music, the canyon with the sun hitting the higher points of pink and yellow, and the congregation held in blue shadow. When the solo was over, the boy remained like a statue, trumpet in hand. The audience waited, expecting another solo; the closing prayer was said; the service was over; but neither the people nor the boy moved, and absolute silence reigned. Finally a man broke the spell by calling to the boy, who still did not answer nor move. As if released by the sound of the voice, the people started to move back to their cars. Three men climbed up and brought the trumpeter down the cliff; he had frozen with fear as he looked down the steep narrow path, his only way back to the valley floor.

The moonshiners are now only a memory, but the sunrisers have grown into many hundreds, with trailers and campers pulling into Red Rock Canyon days early for a good parking spot. The canyon becomes an immense camping city, with many persons watching and listening to the services in the warmth of their vehicles, not braving the chill March or April sunrise hour. Easter Services Association, a cooperative sponsorship of East Kern County churches, provides the services where 40 years ago 10 heads bowed in the first Red Rock Canyon service as the sun's early rays slid down the colorful cliffs.

Bruce "Curley" Carroll lived in Goler Canyon in the 1930s. *(Author's collection)*

Originally Randsburg Sundries was Gunderson and Jones Drug Store, then Hankammer Drug Store. For years the post office sign was on two buildings although the one to the right had long been vacant. The Rinaldi Market building to the right housed the old Randsburg Drug Store of 1898. *(Author's collection)*

21

Today

The Rand area continues to wait for the next boom, whether it be mining, tourist, or something yet to be discovered. Tourists are on the increase, but the end result is a far cry from the million dollar business that was mining.

Retired persons moving into the long-abandoned buildings enjoy the challenge of modernizing the old structures. Newcomers join the old-timers in the struggle for the continued existence of the small communities that tie the area to its illustrious past. The lack of water, a paramount problem in the early mine camps, has been solved for the mountain communities of Randsburg, Johannesburg, and Red Mountain. Through community action a water district was formed, and after what seemed like endless investigations and studies, money became available and a deep well was developed in the valley between Randsburg and Garlock to end a water shortage that dated back to the 1890s in Rand Camp.

Throughout the area weekend businesses come and go, but a few of the long-established ones hold on with determination. In Randsburg the same adobe building has housed the post office for many years and for even more years the building next door has served with merchandise of all kinds. Originally the Gunderson-Jones Drug Store, and later Hankammer's Randsburg Sundries, it has continued very much the same under the management of George and Constance Post. The store's ancient mirror-backed fountain is still a favorite place for the old fashioned ice cream dishes, the Gondola, and fruit-flavored phosphates.

Up the street, Olga Guyett's Joint carries on in the old McMurry's night spot where verbal gold is mined over a glass just as in the old days. Farther up the hill the Randsburg branch of the County Museum gives weekend visitors a view of minerals found and equipment used during the boom mining days. On the same side of the street is the fire station, operated by one man whose multiple jobs include not only fires but information, weather reports, tests for radioactivity, and life saving in an area that covers approximately 24,000 desert acres, mines, mine camps, and communities from Cantil to Red Mountain.

Around the mountain from Randsburg, Johannesburg too, has silent, empty buildings. Along Highway 395 the historic St. Charles Hotel no longer serves the public, the Miner's Trading Post is vacant, and the new, short-lived Laskey Market waits with closed doors. To the traveling public the post office, two filling stations, and the long-established Howard's Cafe, are all that appear to be in operation, but the large all-purpose community building and school are beehives of activity for the youth and adults of the area.

At Red Mountain the homes are owned by the residents and the land is in a state of change from mine claims to private ownership; the old mine frames come right down into the backyard as a constant reminder of past glory. The post office, the Owl Garage, and Owl night spot welcome the visiting public as does the Silver Dollar, last token of the mine camp's roaring past.

In spite of outward appearances there is more life than meets the eye. Behind the seemingly deserted communities, mining, development, and speculation go on. The Atolia tungsten field has mines once more under lease, and machinery for a new method of processing brightens the scene. In the town site of Red Mountain, the Kelly Mine, one-time rich producer, has been thoroughly examined recently with an eye to its potential for open pit mining. At Johannesburg the privately owned airstrip has been reactivated and the renowned King Solomon Mine is periodically checked for future development. On the hill overlooking Randsburg, the Big Butte Mill is being reactivated. With the rise in the price of gold, ore stockpiled years ago is now being processed by Bert Wegman, one of southern California's last experienced gold mill operators.

To the north the Goler Canyon area is seeing newcomers building weekend homes and trying their hand at prospecting or dry washing the old diggings where they hope the old-timers missed a bonanza. Garlock, now a historic town site, is a ghost town of old ruins amid new structures, with a land development program in process. The once free-flowing Mesquite Springs have retreated into the mountain, putting forth only a trickle for birds and small animals.

In the process of modernizing, Saltdale has cleared out all the old buildings and the towering wood mill that was a landmark along the dry lake. Fewer men are needed now to transform the lake's mineral liquor into commercial salt, a product for defogging helicopters, and a material for deicing winter roads of the High Sierra.

Cantil still has a school and until recently had a post office; the retired postmaster, Martin Engel, last of the old-time gold assayers, continues to run ore samples for present-day prospectors. The nearly flat desert around Cantil is fast being converted to ranches raising feed, horses, sheep, and cattle. Rancho Seco, the only sizable settlement in this ranch country, has developed fine homes after waging a battle against shifting sand and flash flooding.

Thousands of acres are now a "multiple use" zone, embracing a civilian rocket research firing range, motorcycle runs, dune buggy meets, jeep trips, camper and trailer club outings, rock clubs, treasure hunters, bottle diggers, gold prospectors, and some persons just trying to escape the city. A search for space and quiet becomes less and less rewarding and the proverbial desert silence is nonexistent. Planes zoom overhead and sonic booms are not unusual as Edwards Air Force Base and the Naval Weapons Center at China Lake go into action. Competing with sounds aloft, the motorcyclist races his motor, and four-wheel drive vehicles add a roar as those in search of recreation claw at earth and plant.

On the day before Christmas, 1969, an intense fire enveloped the idle Yellow Aster Mine's 100-stamp-mill building. Firemen and volunteers battled the blaze ten hours in high winds that carried burning embers and threatened the town of

Randsburg. Old-timers and new residents alike stood watching in disbelief as pointless vandalism turned the last visual monument to the gold era into a twisted heap of metal, massive charred timbers, and ashes.

The few remaining old-timers wistfully hope that mining will make a comeback; young men stand ready to try their hand at full-time mining if the price rises to give a reasonable wage. The very young race along the streets where once the long teams pulled the creaking, groaning freight wagons, where dust rose up around their feet, and the thud of gold mills was background music.

BIBLIOGRAPHY

The sources upon which this book is based are too extensive to list completely. Material used included books, periodicals, manuscripts, government documents, newspapers, pamphlets, scrapbooks, old payrolls, store records, letters, and many personal interviews.

BOOKS

Bailey, Paul. *Walkara, Hawk of the Mountains.* Los Angeles, Westernlore Press, 1954.

Bailey, Richard C. *Explorations in Kern.* Bakersfield, Calif., Kern County Historical Society, 1959.

Bancroft, Herbert H. *Bancroft's History of California.* Vol. VII, Vol. 24, San Francisco, History Publishing Company, 1890.

Beebe, Lucius. *Mixed Train Daily.* New York, Dutton and Company, 1953.

Boyd, William H. *A Centennial Bibliography on the History of Kern County.* Bakersfield, Calif., Kern County Historical Society, 1966.

Brace, Charles L. *The New West.* New York, Press of Wynkoop and Hollenback, 1869.

Bradley, W. C. *Mines and Mineral Resources.* San Francisco, California State Mining Bureau, 1915.

Burmeister, Eugene. *Early Days in Kern.* Bakersfield, Calif., Cardon House, 1963.

California Interstate Telephone. *Romantic Heritage of Upper Mojave Desert.* Victorville, Calif., 1961.

Chalfant, W. A. *The Story of Inyo.* Bishop, Calif., Chalfant Press, 1933.

Comfort, Herbert G. *Where Rolls the Kern.* Moorpark, Calif., Enterprise Press, 1934.

Dagget, Stewart. *History of Southern Pacific.* Undated booklet.

Dibblee, T. W., Jr. *Geology of the Saltdale Quadrangle, California.* San Francisco, Department of Natural Resources, 1952.

Doctor, Joseph E. *Shotguns on Sunday.* Los Angeles, Westernlore Press, 1958.

Edwards, Elza I. *Freeman, Stage Stop on the Mojave.* Glendale, Calif., La Siesta Press, 1964.

Elliott, W. W. *Illustrated History of San Bernadino County.* California, 1882.

Fenton, C. L., and M. A. Fenton. *The Fossil Book.* New York, Doubleday and Company, 1958.

Guddle, Erwin G. *California Place Names.* Berkeley, Calif., University of California Press, 1965.

Harrington, Mark R. *Pinto Site.* Los Angeles, Southwest Museum Paper #17, 1957.

Heizer, R. F., and M. A. Whipple. *The California Indians.* Berkeley, Calif., University of California Press, 1951.

Jaeger, Edmund C. *The California Deserts.* Stanford, Calif., Stanford University Press, 1965.

McGroarty, John S. *California.* Los Angeles, Grafton Publishing Company, 1924.

Manly, William. *Death Valley in '49.* Santa Barbara, Calif., Wallace Hebberd Press, 1929.

__________ . *Jayhawker's Oath.* Los Angeles, Warren F. Lewis, 1949.

"Men of Achievement in the Great Southwest," *Los Angeles Times,* 1904.

Miller, Thelma B. *History of Kern County, California.* Chicago, S. J. Clark Publishers, 1929.

Morgan, Wallace M. *History of Kern County.* Los Angeles, Historic Records Company, 1914.

Nadeau, Remi. *City Makers.* Los Angeles, Trans-Anglo Books, 1965.
Newmark, Harris. *Sixty Years in California.* New York, The Knickerbocker Press, 1926.
Peirson, Emma. *Kern's Desert.* Bakersfield, Calif., Kern County Historical Society, 1956.
Rider, Fremont. *Rider's California.* New York, Macmillan Company, 1925.
Rodgers, Glendon J. *Historic Kern County.* Bakersfield, Calif., Kern County Historical Society, 1948.
__________ *Kern County is our Home.* Bakersfield, Calif., Merchants Printing and Lithographing Company, 1950.
Shelton, John S. *Geology Illustrated.* San Francisco, W. H. Freeman Company, 1966.
Shinn, Charles, H. *Mining Camps.* New York, Alfred A. Knopf, 1948.
Smeaton, Charles J. *California Desert Trails.* Massachusetts, 1919.
"Southern California, Land of Hearts Desire" *Los Angeles Morning Herald*, 1911.
Troxol, B. W., and P. K. Morton. *Mine and Mineral Resources of Kern County.* San Francisco, California Division of Mines, 1962.
Walter, William W. *The Great Understander.* Aurora, Ill., William W. Walter, 1931.
Whiting, Perry. *Perry.* Los Angeles, Smith-Barnes Corp. 1930.
Wilson, Neill C. *Silver Stampede.* New York, Macmillan Company, 1937.
Wynn, Marcia. *Desert Bonanza.* Culver City, Calif., N. W. Samelson, 1949.

GOVERNMENT PUBLICATIONS

Bradley, W. C. "Mines and Mineral Resources." California Division of Mines, 1915.
"Geology of the Saltdale Quadrangle of California," California Division of Mines, Bulletin 160, 1952.
"Mines and Mineral Resources of Kern County, California," California Division of Mines, 1962.
Thompson, David. "Mojave Desert Region." U. S. Department of Interior, Paper 578, 1929.
National Archives. Post office records.
"Routes to Desert Watering Places in the Mojave Desert." Department of Interior, Water Supply Paper 490B, 1921.
"Source Material for Use in the Study of Kern County Indians." Kern County School Study Bulletin, 1944.
"Survey Field Notes, Eastern Kern County." Kern County Survey Department, 1853-54. Wanton, J. S. "Indian Tribes of North America." U.S. Government Printing Office, Bulletin 145, 1953.
Wanton, J. S. "Indian Tribes of North America." Washington, D.C., U.S. Government Printing Office, Bulletin 145, 1953.

NEWSPAPERS

Barstow Printer, July 1920
Boston Eagle, 1899
The Bakersfield Californian, July 1936
Desert Mountain News, 1954
Los Angeles Times, 1898
Randsburg Miner, 1900-1932
Randsburg Times, 1930s
Tehachapi News, 1928-1934

INDEX